TARGET
ZERO

TARGET ZERO

ANTHONY RICHES

An Aries Book

First published in the UK in 2022 by Head of Zeus Ltd,
part of Bloomsbury Publishing Plc

9 7 5 3 1 2 4 6 8

A catalogue record for this book is available from
the British Library.

ISBN (HB): 9781801109970
ISBN (XTPB): 9781803281339
ISBN (E): 9781800249004

Typeset by Siliconchips Services Ltd UK

Printed and bound in Great Britain by
CPI Group (UK) Ltd, Croydon CR0 4YY

MIX
Paper from
responsible sources
FSC® C171272

Head of Zeus
First Floor East
5–8 Hardwick Street
London EC1R 4RG

WWW.HEADOFZEUS.COM

For Helen, as always.

Prologue

Somewhere in the back of his mind, Darren Pearce was still playing catch-up.

Not in terms of the ground reality. His team had responded to the scramble call like the pros they were. Following the drills trained into them over several years. Endless preparation for something unlikely ever to become real. Now closing in on the designated point on the map as directed. Not much more than a minute or so to spare, from the sound of the radio chatter.

Sergeant Pearce and his team had been patrolling the airport. Stansted, Ground Side. The duty armed response team. A routine gig for the Essex force's Field Support Unit. Equivalent to the Met's MO19. Heavy-duty armed coppers. Two teams of two. One officer facing forwards, one facing backwards. Patrolling through the holidaymakers and businesspeople. Looking out for the kind of trouble that hardly ever happened. But which could kill dozens if it did.

Still working coppers, of course. Looking amenable. Fielding questions. Giving directions. Swapping bants. While all the time alert for anything out of the ordinary. High focus, energy sapping, but low drama. The sight of a rifle having the tendency to smother drama at birth. The norm in a country still unused to heavily armed police.

Making the job a bit boring, truth be told. Although for a specialist firearms officer boring was a good thing. True, Pearce occasionally dreamt of getting a chance to do the exciting side of the job. But knew that dream could become nightmare in the blink of an eye. 'Be careful what you wish for' stuff.

And then, without warning, an urgent radio message. Turning it all upside down. No warning. No 'this is a drill'. Punters gawping as they followed the emergency protocols. Legging it from the far end of the busy concourse to their carrier.

Driver waiting, engine running. Blue lights already flickering. Going flat out the wrong way round the airport-approach roundabout. Avoiding a wasted mile on the road if they'd taken the legal route. And dispelling any doubts. This was no drill.

Redlining it to the M11. Directed to head north up the motorway. Scything through the traffic as far as the unmarked turn off at Wicken Water. Under the carriageways and back on, southbound. Pearce and his team getting their breath back.

Briefed on the move, as they kitted up. Duty inspector on the radio from the other side of the county. Filling in the detail, sounding as jealous as fuck. Suspected Islamic terrorists heading for London from the north. Routing down the motorway past Stansted. Heading into the target-rich jihadi honeypot of London.

One vehicle, four suspects. Pearce's team's mission being to stop and detain. Standard rules of engagement. Arrest if possible. Use of firearms authorised if necessary to preserve life.

Pearce watched his troops as they kitted up. His own preparation so practised as to be on autopilot. Securing Kevlar 'Fritz' helmets, ballistic goggles ready for use. Strapping on heavy armour plate carriers over black boiler suits. Swaying in the moving vehicle, but getting it done quickly and without fuss.

Belt kit going on last. Everything needed for a variety of circumstances. Spare mags. Trauma kit. Restraints. Smoke grenades and flash-bangs. He talked them through one last weapon check. Loaded mags, rounds chambered, safeties off. Condition one: fingers off triggers.

All four of them equipped with Heckler & Koch G36C carbines; 5.56mm. Serious military firepower. Everything falling into place. The way they'd rehearsed it time after time. Pearce's team was a machine. But in the back of his mind, something was still nagging.

The van was slowing. A motorway spec X5 parked on the hard shoulder. X marks the spot, his mouthiest officer joked. Her face pale. Fight or flight kicking in. The driver pulled in, in front of the BMW. Leaving a fifty-metre gap between the two vehicles.

'Listen in!'

An army cliché, he knew. One they constantly ripped the piss out of him for. But old habits die hard. And the adrenaline would be fizzing in their veins. Commands would have to be crisp and clear.

'We debus! We stay in the cover of the carrier! Keep it low, avoid being seen! The Road Policing boys will deploy the Stinger! Then *we* do the heavy lifting!'

Which got him some smiles. The radio interrupted before he could continue.

'Foxtrot Sierra Uniform from Whiskey Two Four. We're the Romeo Papa crew at your six. Target vehicle sighted, estimate forty-five seconds to contact.'

Pearce's response was immediate.

'Roger, Two Four. Call out Stinger release, with a countdown please. All stations, clear channel, clear channel!'

That got a series of clicks in response. Nobody was going to talk over an operation like this. A few seconds remaining to finish his orders.

'You know the drill! If you have to shoot, go centre-mass! Keep shooting until they go down and do *not* stop shooting to assess!' Nodded to his most experienced officer. 'Richie, you're lead! I'll be right behind you, then Jenny, then Karl! Once we're out of cover we go echelon right! I want maximum firepower if it kicks off! Right, let's go!'

Richie pulled the carrier's side door open. Climbed out and moved forward. Staying low and close to the vehicle's side. Crouching in the cover of the van's left front wing.

'Whiskey Two Four. Stinger in thirty seconds.'

The Road Policing officer sounding calm. As if there was nothing more exciting than a sit-down with a cuppa in the offing. Pearce mentally doffed his cap to the man.

Richie crouched by the front offside lights. Put a padded knee onto the tarmac. Raised a hand and shouted the expected statement of preparedness.

'*Ready!*'

Pearce echoed the shout. Looked back at the other two as they replied. Both on one knee behind him. Weapons pointing at the verge. A good team. No, a great team. He knew they'd do him proud.

'Stinger in twenty seconds.'

Still something at the back of Pearce's mind. Something he'd lost in the rush to embus and kit up. Richie looking back at him, waiting for the go order.

'Stinger in ten, nine, eight...'

And Pearce realised what it was that was bothering him.

'Six, five, four...'

Too late to ask the question now. He exchanged a last-minute nod of mutual respect with Richie.

'... two, one... Stinger deployed!'

The suspect vehicle whipped past them with a howl of protesting rubber. Tyres already deflating. Embedded hollow spikes allowing their air to escape in a rush. The driver's foot hard on the brakes. Barely controlling the car as it skidded into the outside lane. Coming to a halt forty metres past the carrier.

'Go! Go! Go!'

Richie led them out onto the empty motorway. Carbine raised. Pearce a pace behind him and to the right. Carbine held on target with one hand. The other keying his mike back to the control room in Chelmsford.

'Control, FSU. Suspect vehicle stopped. Interrogatory: is there an explosives risk? Urgent response please!'

Still moving forwards, hearing the mosquito whine of a drone in the air behind them. Pearce gestured with two fingers for the other two to echelon further out to his right. Wanting more than one rifle on target if the suspected extremists had the means of going loud.

'FSU, Control—'

The doors of the battered-looking Vauxhall opened. Four men in the act of getting out. And anything that control had to say was suddenly superfluous. Way less important than the situation to hand.

'*Armed police! Down on the ground!*'

Richie amping up the volume to a roar. As trained. Intimidate the target hard enough and you might not have to shoot them. Or so the theory went. Jenny and Karl were shouting too. Still advancing. Pearce drew breath to reinforce the bellowed commands. Just as a single word from control caught his attention. Burning through the cognitive chaff that was overloading his thought processes.

'...EOD is inbound, ETA three zero.'

EOD. Explosive Ordnance Disposal. *Shit.*

He stared at the suspects. Close enough to see their faces. One just scared. Not really realising the depth of the shit he'd stepped into until this minute. Another frankly terrified. And looking like he'd already pissed himself, or worse. One with the beatific smile of a true believer. Which was heart-stopping enough in itself. Indicating an expectation of imminent martyrdom.

And the fourth... Pearce felt the adrenaline surge hit him. His animal brain made a prompt decision to kill the fourth man. Jumping straight to that conclusion well before his conscious mind caught up.

The fourth man was grinning hard enough that his face was almost a rictus. Teeth bared. An expression that shouted intent. With a small black box in his right hand. The subject of his amusement. Pearce made the connection in an instant. EOD. A black control box. And a grinning extremist. Joined the dots.

The grin turned into a snarl even as Pearce put his hand back on the carbine and lined up the shot. Putting the death dot on the other man's forehead. His finger onto the trigger. No hesitation, just action.

He fired, the gentlest squeeze of his index finger, unleashing the terrible power of the 5.56mm round chambered and waiting.

Just one tenth of a second too late. Sending the bullet straight into the incandescent wall of flame that incinerated him at the same instant.

1

Friday evening. Monken Park Boxing Studio. Fashionably bare brick walls, tiered rows of punch bags and speedballs.

Outside, rain on the streets, unseasonably cold for late May. Commuters scurrying home with their hoods up. Some still wearing masks, as much for warmth as protection. Eager for central heating and oven-ready. How was your day, or just telly's solitary embrace.

Inside, warm and aromatic. The smell of sweat, liniment and grim determination. Half a dozen paying punters pounding bags. A couple more working with their personal trainers. And in the ring, two men. Mickey well past his prime in fight terms. The kid yet to reach it. But Mickey knew the kid had him in deep trouble the moment the fight started. Grizzler gestured for the two men to get to it. And the shit promptly hit the fan.

To be fair, it wasn't like Mickey hadn't been warned. The kid had strolled across the ring towards him nonchalantly enough. Gloved hands down by his hips. And a knowing smile that set his instincts jangling. Leaned in, as if to say hi and thanks for offering to spar. And delivered the good news through his Irish flag gum-shield. Flat stare, flat monotone.

'My da says you're filth. Says you smacked him, good and hard, one Saturday night. On the High Street back in the nineties. Says you knocked him spark out. So I'm gunna make you my bitch, right? I'm done with you, they'll be doing the ten-bell salute.'

Slight overkill with the final ten-count thing. But effective enough as threats went. He'd looked Mickey up and down with the expert eye of a butcher considering a side of beef. Cleaver in hand.

Seeing, Mickey knew all too well, a slightly creased and dog-eared former copper in his forties. Gym-toned, minimal body fat, bullet-hole pucker in his right abdominals. Muscles, poise, and the scar to match. Fit, and known to be double hard. But still well past his sell-by for the ring. Ready to be battered, likely to be deep-fried, by the right fighter. Kid gave Mickey one last disparaging look. Turned and walked back to his corner.

Kid was this year's bright prospect. Potential superstar in the making. Not wholesome like Joshua. Not a dancing joker like Fury. A wild-eyed animal of a fighter. Known to idolise Conor McGregor and his ilk. And a product of the club Mickey had been boxing at for the last three decades. Tipped to be at least a finalist in his weight class at Amateur level inside the year. Despite only being seventeen. Runner-up for the National Youth Championships a month before. Unlucky to have lost to a split decision. The fight fraternity exchanging knowing glances at that result. And champion at Junior and Youth levels for the past three years.

In fight terms, the kid was red-hot. Fast feet. A granite jaw. And the full deck of offensive options. A potential world champion, with the right management. Which management

8

was sitting ringside. Accompanied by suitably hench minders. Watching his new boy work out. Chewing his way through a disappointing succession of mobile punchbags. With the kid's da sitting next to him. Looking daggers at Mickey.

Not that Mickey knew him from Adam. But then he'd decked a few lairy geezers on the High Street in the nineties. In his first years as a beat copper. Back in the days when he'd been all piss and vinegar. And when the odd sly dig of a difficult punter had still been acceptable. So go figure. And now the kid was coming for revenge. Intent on reducing Mickey to a lumpy paste.

Grizzler had wandered across, pre-fight. Raising an eyebrow. Good instincts, Grizzler. Probably picking up the vibe from the kid's father.

'You owight, Mickey son? Was good of you to offer up when he put his last partner to sleep early, but the little bastard's like a junkyard dog when he gets started. No shame in changing your mind, eh?'

Last chance to bail on the whole stupid idea. Make his excuses and climb back out through the ropes. Rather than be carried out through them.

'I'm fine, mate. Happy to help the kid out.' Mickey almost unable to believe his own bravado. Other descriptive nouns being available. Grinned at the gnarled old trainer through his gum shield. 'What could possibly go wrong?'

'D'you want a *list*?' A shrug. 'It's your funeral son. Just don't say you wasn't warned. And watch his left. He uses the right to intimidate, but the left hook's the killer.'

And walked back to the centre of the ring. Getting ready to usher them both forward. Leaving Mickey to ponder just what the fuck he'd got himself into. Mickey Bale, ex-copper, ex-protection officer, currently at a loose end. Moderately respected around his own manor. Yes, he was still a copper. Once a copper always a copper, right? But he was always straight, innit? Which was what Mickey liked to think was the consensus of opinion. Knowing he was probably being optimistic.

He'd recently returned from abroad. A year away, soaking up the sun. And getting over events that had made him an ex, three times over. Ex-copper. Ex-Met protection officer. Ex-husband. Fit as fuck from all the training he'd indulged in, true enough. But in boxing terms as ring-rusty as a fifty-year-old Lancia.

The bell rang, and Grizzler pointed to the centre of the ring. And Mickey walked out to take his twenty-year-deferred spanking.

The first ten seconds confirmed the likely degree of punishment. And took Mickey straight to full-on brown alert. Kid came in fast and hard. Throwing combinations from the get-go. And nothing trivial. Inside the first minute he'd landed a half a dozen tidy body shots. The professional boxer's go-to punch of choice for the early stages of any encounter. Unless of course an opponent left their chin hanging out. Only so much punishment a man can take before it starts to hurt.

Mickey, wearing the full muscle jacket, rode the blows easily enough. But knew he was taking damage. Imagining his health bar starting to dwindle. Tried an experimental jab, leaning in to get the shot away. And missed. The little bastard's lightning reflexes taking him around the punch. Like trying to hit smoke. Took a short right to the face in return. The punch's raw power staggering him back, blinking. Had the kid's dad, manager and hangers-on roaring approval. Kill the filth, nail the prick, yadda yadda yadda.

Mickey took strength from their hatred. Raised his gloved fists and went back to basics. No way he was going to last if he stood up and fought. No way he could box the kid for three rounds. Fell back on the two most simple defensive tools he had to hand. Taking the beating on his hands. And running away.

He blocked and shuffled backwards in a circle. Feeling no shame. And giving no shits for the abuse he was taking from the audience. Made it to the end of the round intact. Went back to his corner with an inward sigh of relief. Rinsed his mouth out. Spat in the bucket Grizzler offered him.

No blood in the water.

Yet.

'He's got you pinned, Mickey son. Got any ideas?'

Mickey looked up at the ancient trainer. Who'd already been well middle-aged thirty-odd years before. Mickey's stepdad Terry having suggested the club. And delivered him into the arms of redemption. A new purpose in life. And the ability to fight back.

Grizzler, named for a one-off shedding of tears. Prompted by an alumni of the gym's brief success on the pro circuit. Before Mickey's time, but still the stuff of legends. And Mickey, still breathing hard, knew exactly what he had left in his locker. Knowing that he had to restrain himself from delving into his new bag of martial arts tricks. It was the Marquess of Queensberry or bust.

'Thought I might run away for the rest of the fight. Give him some practice unpicking locks.'

A grim headshake.

'You ain't got the wind. He ain't got the patience. Think again, son. Take another three minutes, eh?'

Thanks, Mickey thought. Nodding grimly and standing back up. The usual Grizzler opinion. Terse, pithy, invariably correct. It was a quirk of fate that the old man had never had a Lewis or a Benn walk through his doors. Leaving the club under-regarded by the cognoscenti. But like Arsene, Grizzler knew.

Kid came out under full sail. Determined look on his face. Trainer in his ear for a solid ten seconds before he stood up. Probably telling him to close out the fight quick. Telling Mickey that he'd have to do better than be a bag on legs. If he wanted to survive the next one hundred and eighty seconds, that was.

So Mickey fell back on the endless repetitions of his youth. Moving up from running away to moving sideways. Not exactly dancing, not at his age. But up on his toes. Taking random lateral steps to throw the kid's aim off. And taking his blocking game up a notch. No longer just letting the kid smack his hands to a pulp. Actually parrying. And trying the odd shoulder roll.

Smiling inwardly as the kid's impatience started to show. Took another shot to his face when he rolled instead of parrying. The kid second-guessing him with instinct beyond his years. But found himself able to grin back, giving it the old 'that all you got?' eyebrows. If round one had been a straight beating, round two was still a points loss. But better. Grizzler held the bucket up. Grimacing at the blood washing around against the white plastic.

'He's gonna have you this round. Coming out with the big one ready to throw. *You* be ready, eh?'

And walked away. Mickey shook his head in dark

amusement. Gnomic being a word specifically invented for Grizzler, he reckoned. But he knew exactly what the old man was telling him. Time to die, so to speak. Unless he had something more to offer.

In a boxing movie the kid would have said something at the top of the third. Something spiteful, or goading. This being real life he just gave Mickey the stare and got down to business. End game stuff. Storming in with a raging flurry of combinations. One eye constantly on teeing Mickey up for the big one. Just as advertised. Manoeuvring Mickey round the ring with one thought. To put that golden punch through his crumbling defences. Deliver on his promise to the pack baying for blood behind him.

Mickey notched his defensive game up again. Slipping to left and right. Channelling what was left of his energy to proper boxing. The years momentarily washed away on a last-gasp wave of adrenaline. Even the gym punters were watching now. One or two nodding respect. Genuinely amazed that the old guy had made it to the third with more than just rope-a-dope left in him.

And then the kid's da made the fatal mistake. Inflamed at the thought that his one-time persecutor might yet escape punishment. Brushed off the kid's manager and jumped to his feet.

'Fokkin' *hit* him!'

The roar enough to give Mickey a fleeting moment of opportunity. The kid's attention spilt, for one beautiful, yawning instant. Caught between opponent and instruction. Better yet, caught extended. Right arm out feeling for the chance to unload his left. A chance, the only chance of the fight, on offer. And Mickey took it.

Stepped in. And counter-punched properly for the first time in over five hundred seconds of fighting. Fired his trademark right hook in from low. The punch finding the kid wide open. And liking what it found. Inch perfect, straight to the on-off switch. Lights out. Kid went down like the proverbial sack of shit. Eyes rolling to show the whites, mouth falling open to spill his gum shield.

Mickey turned away as he hit the canvas. Winked at a simultaneously disgusted and secretly delighted Grizzler. Grinned down at the kid's da. Raising his gloves and making the universal gesture for 'you want some?' Sending a message, loud and clear. Have some class. Step up or shut up. Watching as the famous manager's goons went round the gym. Confiscating phones and deleting camera files. Mickey pretty sure there was no way they'd get them all. Thinking uncrimp that, motherfuckers. Mickey Bale most definitely back in town.

2

James Cavendish had seen his future in the smooth, emotionless features of his regiment's colonel. His commanding officer having come to visit him in his hospital bed. A bed to which James was confined, having been shot through the shoulder.

His military career had been illustrious, fair to say. But he wasn't fated to reach senior rank. A previous colonel had spelled it out all too clearly over coffee and biscuits years before. Telling James that he'd reached the limit of his realistic ambition. That he'd never go any higher up the chain of command than major. Ever.

His service record, of course, was near perfect. Secondment to the Special Reconnaissance Regiment, as a captain. Two years hunting the Real IRA. And with excellent results. Their organisation compromised, weapons caches jarked, would-be terrorists tracked. An exchange tour with the 75th Ranger Regiment at Fort Benning. Again, a roaring success. A gushing shower of superlatives from across the water. And unquestioned excellence on domestic duties, both ceremonial and in the field.

James was, all had agreed, the quintessential soldier. But

with a face that simply didn't fit, it seemed. Because James was, in the army's eyes, the archetypal dangerous radical. Having made the mistake of expressing his frustration with outdated weapons and tactics. Naively taking the tuition staff at their word when given the chance to speak freely at staff college.

And so he had heeded the writing on the wall. Hard not to, given that it was written in his military career's lifeblood. And had accepted his superior's sympathetic assistance with carefully phrased gratitude. Having been promised a squadron command position with some *very* special forces. *If* he could pass selection.

Which was a *very* big if. But with the potential reward of service with Them. The Blades. The Men in Black. Official title: the 22nd Special Air Service. His colonel being tight with their colonel, a fellow Sandhurst classmate twenty years before.

He had literally staggered though selection. At the age where even athletes' fitness was on borrowed time. Perpetually on his last legs, his conscious mind screaming at him to quit. But never allowing himself to give up. Or lose his temper with a succession of sickeners. Which he suspected – correctly – had been designed to bomb out the likely to be useless Rupert.

The lorry that drove away as he got within touching distance. After tabbing twenty-five miles across a freezing and rain-blasted moor.

Opening the food drop that he'd been dreaming of for the last twelve hours. And finding a live rabbit and no means of cooking it.

And many, many more. All swiftly dismissed as mere inconvenience. And dealt with, smiling... or at least grimacing… all the way. Awarded the fabled sand-coloured beret, he'd immersed himself in the minutiae of special forces. And enjoyed the best years of his military career.

Without a meaningful career to return to, he'd lasted for an improbable five years with the Regiment. Twice the average, for an officer. Taking advantage of the need for solid officers. The Regiment being in action at a furious operational tempo across the Middle East. A glorious swan song. Albeit ingloriously ended.

Ended because he had started to suspect some overzealous prosecution of 'the enemy'. Blurring the boundaries as to the legitimate definition of that term. And leading to the toleration of needless civilian deaths. His tour promptly terminated when he had voiced his suspicions to his superior.

And so James had found himself on the outside looking in within twenty-four hours. RTU. Return To Unit. Not the result he had hoped for, but in any case somewhat overdue.

Happy with his personal ethics, he'd reported back to Wellington Barracks. Conscience clear. Ready to accept pretty much whatever the Guards had for him to do. Which had turned out to be nothing whatsoever. He was completely surplus to requirements. Although there was one job on offer. Just not with the Grenadiers. The advice being that it was 'a bloody important role'. And that 'I'd snap it up if I were you'.

The bloody important role being that of the defence secretary's new private secretary. Effectively a civil servant in uniform. Expected to be the army's eyes and ears in the Secretary of State's office. Making James's known abilities and absolute discretion perfect for the role.

But it was a role that had ended abruptly. Leaving James seriously wounded. Flat on his back with a serious bullet wound. In a road littered with dead and dying police officers. Hors de combat, but satisfied he'd done his duty one last time. Albeit at the cost of stopping a rifle bullet that had trashed his right shoulder. A supersonic red-hot poker blasted through the complex bio-mechanical structure of bone, muscle and cartilage. Trashing whatever was left of his military career in the same instant. Putting him in a hospital bed, scheduled for a replacement joint operation.

Which was what had led the Regiment's current colonel to visit him. His latest boss unsentimental to the point of brutality. Disappointing, but unsurprising. Given he knew the man all too well from a tour in Afghanistan years before.

'You're finished I'm afraid, Major Cavendish. No longer fit for duty as an infantry officer. You'll get a decent disability pension though, and the medal will help you attract employers, I'd imagine.'

He'd sniffed, flicking a glance down to his Distinguished Service Order ribbon. Awarded for 'exceptional leadership in the Afghan conflict'. His disdain rooted, James supposed, in the fact that James's George Cross, awarded at the bedside the previous day, wasn't an operational medal. More of a gong for 'civilians' in the colonel's blinkered little world. Even if it did rank alongside the Victoria Cross.

He'd waited for his newly minted superior's next statement with a mix of amusement and bitter resentment. Wondering just how good a punch he could get in from a recumbent position with his left hand.

'Although there might just be light at the end of the tunnel for you. I received this yesterday.'

James had read the proffered hard-copy letter. Realising that it was likely to be the only reason for colonel number three's personal visit. An opportunity to get a loose end tied off. Make room for a promotion from captain for some lucky man. A swift and efficient removal of a career blocker. Just the way colonel number one, still watching from division, would need to see it done.

The Security Service, it seemed, had an opening for a man like James. Having got to know him better in the course of his brief tour looking after the defence secretary. Deeming him to have known positive qualities, long military experience, et cetera et cetera. And his time with the Regiment had been the crux of the matter, apparently. The colonel was predictably disparaging. Keen to see James go. Just not willing to sugar the pill. Too bound up in his self-evident narcissism.

'Seems they want someone with just enough savvy to stop the Sports and Social going off the rails when they do their dirty little jobs. Shooting the wrong person. Misreading the map and ending up on the wrong side of the Scottish border. That sort of thing. Shouldn't be too difficult, even without a functional right arm, eh?'

James had pondered. On the upside: continued pay, pension contributions and London allowance. Most importantly, continued service. And on the downside: well... nothing that he could see from a hospital bed. With a shoulder that was never going to be robust enough for soldiering again.

He contemplated the matter, and the colonel's offhand delivery of the lifeline, for a second or so. And made a swift decision. His intelligence steering him around the urge to take the other man by the throat. But declaring him weapons-free when it came to verbal assault.

Swivelling painfully, he'd put the letter down on his bedside cabinet. On the other side of the bed. Rather than returning it to colonel three's waiting hand. Retaking the power in their non-relationship.

'I'll take it from here, thank you. I intend to accept the offer. You can report back that I'm no longer the Regiment's problem. Or division's. Job done, eh? But before you go, I'll let you in on a secret.'

The colonel leaned a little closer. The look of suspicion on his face well merited.

'Go on.'

'You know those bouts of dysentery you were so prone to in Helmand, back in 2012?' The other man nodded. 'Your after-lunch orange habit wasn't quite as blameless as you might have thought, in that respect.'

Gave it a moment for the colonel to process what he was saying.

'What do you—'

'The oranges, Rupert, were spiked.' James dropping both formality and any last pretence at respect. 'You were so fucking dangerous with a map and a radio that we nobbled you. Injected your oranges with untreated water every time it seemed you were on the mend.'

'You impertinent—'

James had raised a hand. 'Shouting wouldn't be wise. They'll show you the door for upsetting the patient. And besides, the joke was on us, wasn't it? You being casevaced to the UK was in the plan. But you getting a Distinguished Service Order for your "inspired leadership in the field" definitely wasn't. Ironic too, when your nickname among the troops was Ever Ready.'

The colonel's face hardened. Knowing all too well that the nickname was an insult along the lines of 'never goes out'. In this case, outside the wire that demarcated the fabled rear-echelon mother fucker from fighting soldiers.

James delivered the coup de grâce. 'I didn't ever use the term myself, of course. Didn't consider it as sufficiently reflective of just how much of an incompetent you really are. And now I think I need a sleep. All this excitement has completely worn me out. Sir.'

Sir with the archetypal silent c, as both men knew. James severing his last lifeline to the Regiment in that moment. With regret, but also with enough élan that he doubted the man would ever meet his eye at a regimental association gathering. And just like that, James had gone from soldier to spook.

3

Two days on from Mickey's unexpected victory. His aches and pains still aching and painful. Face mercifully unmarked, apart from a cut mouth. Body covered in purpling bruises. But nothing that wasn't immediately salved by the memory of his unlikely victory. And the look on the kid's face when he'd come round from his brief sleep.

The papers and TV were still screaming headlines about the latest terrorist outrage. The M11 still closed in Essex. Causing chaos in itself, but overshadowed by the cause. Five police officers dead. Two more hospitalised. One of them not expected to live. And four terrorists, of course. Not that anyone other than their devastated families cared about them. Very little about what had happened was clear. At least not from what the press knew. Or from the wild speculation that was filling the news vacuum.

An anti-terrorist operation gone wrong. An explosion loud enough to have been heard at Stansted being all the media had. Other than the unavoidable facts from the Essex constabulary and a COBR press release. Any witnesses having been persuaded to keep their mouths shut. One way or another.

All of which had prompted the inevitable storm of speculation from all the usual suspects. Special forces veterans with books to sell. Under-informed and overexcited talking heads. And the occasional sober expert commentator. The latter doing their best to unpick the story without sensationalising. Mostly failing, in a confected gale of breathless twenty-four-hour repetition and embroidery.

The consensus of opinion being that the terrorists had somehow raised their game again. That the country was lucky to have such selfless guardians of law and order. With the obvious and inescapable conclusion being drawn. That the Security Service must have been asleep on the job. And that *something*, obviously, needed to be done.

Something being where the consensus ended. And where Mickey switched off. Knowing that opinions were indeed as common as arseholes. A cliché that applied in spades when it came to bombs and dead coppers.

Kev being due at six, Mickey showered and dressed. His first meeting with an old friend since getting back. Kev having made contact the day before. A call, out of the blue. *Now you're back, let's have a bit of a chat, eh?* Likely to be an interrogation of sorts, Mickey suspected.

The doorbell rang at two minutes to. Kev waiting on the landing. Kitted up in the full urban biker rig. Flip-front Shoei helmet, front flipped. Head to foot Alpinestars leather, like a racetrack refugee. Elbows, knees, chest and back all bulging with protection. Grinning at Mickey like the old mate he was. Mickey's mentor, back in the day. Back when Mickey was still a wet-behind-the-ears street copper. Their relationship rekindled, thanks to the pitiless flail of events a year before. Kev and Mickey sharing a good-sized dose of survivor guilt.

'Fuck me, it's the creature from the black lagoon.'

Which got him a pitying look.

'Don't give up the day job, eh son?' Kev never knowingly under-retaliating when it came to swapping endearments. 'And fuck you very much too. Indoor management doesn't let me even *sit* on a bike without enough Kevlar to stop nine mil. New rules, since you know what. Got the kettle on?'

The retired policeman pushed past him into the small flat, shedding kit as he went. Helmet and jacket dropped on the end of the sofa, revealing a lumpy vest beneath.

'What the hell is that thing you're wearing?'

Kev unclipped the harness and tossed it over.

'Chest and back airbag. The latest thing for riders who want to live when they collide with the likes of you, in your big armoured cars. Wish I'd been wearing it six months ago.' He opened the front of his shirt. 'Might have avoided this.'

A small nub of white scar tissue, nestled in his greying doormat-thick chest hair. An unwanted memento of the last day of his career in the Met. Having been on the verge of retirement from the Special Escort Group. A rosy scenario shattered by an ambush in west London. Suicide bombers, Chechen mercenaries, the full nine yards.

Full-on mayhem with military weaponry. Eleven dead police officers and two dozen wounded civilians. Some of them lucky not to have joined the death toll themselves. And only luck had prevented Mickey and Kev from sharing that fate.

Luck and, of course, James Cavendish. Not for the first time, Mickey wondered where the major was. Cavendish and Mickey not being the types to stay in touch. Their social worlds unlikely ever to collide. Cavendish's milieu being Eton, the Guards and White's of St James's. Mickey's being Monken Park comprehensive, the Met, and the Lamb and Flag on the High Street.

'Tea?'

'Not half. I'm parched.'

Mickey nodded. The decision to make a brew the new decisiveness. Given the absence of anything meaningful to do. He walked through to the tiny kitchen. The kettle already hot. Water preheated, Kev's tea and cake consumption legendary. He flipped the button. Flashed the hot water back to a rolling boil.

'NATO, right? Milk and two?'

'S'right. Proper heaped spoons though, none of your health-kick bollocks.'

Mickey took the teas into the lounge. Kev having occupied the other half of the sofa. The older man sipped his drink gingerly.

'Perfect.' He looked up at Mickey speculatively. 'So...'

'So what?'

Kev raised his eyebrows, evidently unimpressed.

'So what? What do you *think*? Where have you been, you muppet? Last time I saw you was a year ago. And me lying in a hospital bed with a sucking chest wound! The sort of thing that takes somewhat longer to get over than your neat little bullet hole. And by the time I was properly back on my feet you'd fucked off. Leaving your missus on her jack.'

Mickey nodded, knowing Kev's fondness for Roz.

'Kev mate, it was kind of the other way round? She invited me to leave the family home. And went straight to decree nisi without giving me the option.'

Kev snorted disbelief.

'She walked because you were a naughty boy, didn't she?'

Mickey pursed his lips. Knowing that not answering was by far his best policy.

'Depends what you mean by naughty.'

Kev locked on with the stare.

'I don't mean over-the-side naughty. I mean *psycho* naughty. Properly out of your pram from the rumours I'm hearing. *You?* Mickey Bale? Slotting drug dealers and gang bosses? How you walked out from under a thirty stretch is still the subject of intense speculation.'

He upped the power on the stare. Grimacing as he took in Mickey's determined no-comment face.

'Like that, is it? Nothing to say?'

'Better that way. What you don't know...'

'Can't be used in evidence against *you?* Fair enough. Got anything to eat in this hamster cage?'

Mickey rolled his eyes and went to cut a slab of Kev's favourite. Followed by an expectant ex-police sergeant.

'That's not a proper slice... a bit more... perfect.'

He took the proffered plate and returned to his perch. Leaving Mickey to cut a thinner slice and follow him back into the lounge. Where Kev resumed the interrogation through a mouthful of Dundee cake. Then grinned, shark-like, at Mickey's impassive expression.

'You can do the poker face all you like, buddy. I have my sources. And anyway, all that crap about the George Cross hero forced into retirement by his wounds? My arse! I mean look at you!'

He had a point, Mickey was forced to concede. Who was probably as fit as he'd ever been.

'The time away didn't hurt.'

'Too fucking right! Tanned, no gut. Muscles even. You look proper hench! Where you been all this time?'

Mickey sighed. Knowing he might as well get it over with. And tell Kev as much as he wanted him to know about the last year.

'Well I wasn't going to sit in a rental flat, was I? What's there to do here except drink, watch box sets and get fat? So I decided to get away for a bit. Thailand, for the diving.' And, more to the point, the martial arts. He didn't add. Seeing no reason why Kev needed to know about his newly developed skills. 'I'm still not entirely sure what it was made me come back.'

Kev guffawed. 'Diving? At your age?'

Mickey smiled back. 'You'd be surprised. There's old bastards *your* age out in the sea.' Grinned as Kev theatrically put a hand to his chest. Doing the pantomime wounded look like the old pro he was. Touché. 'I did the whole dive dude thing, y'know? Grew a beard, bought some expensive equipment. Rented a place on the beach...' He grinned at the memory. 'Did that lifestyle for a few months. Loved it too. I was even thinking about getting my ears pierced and growing my hair long.'

He sensed it was time to switch the subject.

'So how did you know I was back?'

Kev sniffed disdainfully. 'I've been calling your phone every week for months. Getting a foreign ringtone.'

'Ah, that was you, was it?' Mickey's phone had indeed rung on a routine basis. Usually at night, never for long enough to answer. Leading Mickey to put it on silent after eleven. 'Why didn't you wait for me to answer it?'

The older man shrugged.

'Because I didn't want to talk on the phone, that's why. I wanted to see you.'

'Aw.' Mickey grinned at his friend. Part genuine pleasure, part piss-take. 'And here we are, together again!'

'Yeah all right, there's no need to be a prick. I was worried about you. If I'd known you were busy shagging your way across South-East Asia I'd have...' His eyes widened as Mickey shook his head. 'You're never going to try and tell me you weren't on the pull at any point?'

Mickey shrugged.

'Not trying to tell you. Just telling you.'

'I knew it! I told you you're not done with Roz!'

Mickey shrugged again. 'Like I said. Doesn't matter whether *I'm* done with her or not, does it. Because she's the one in charge of that decision. It's over. I'm just not ready to move on yet.' He sighed, and took a sip of his cooling tea. 'I've been with her half my life, Kev. And I don't actually want anyone else at the moment.'

'Fair enough. You're a good-looking bloke, so I don't think you're going to go short once you put the sign up.'

'The sign?'

'You know. Situation vacant. Open to offers.'

Mickey smiled. 'We'll see. One other question though.'

'Wassat?'

'How did you know I was actually back in London?'

Got a pitying look from Kev.

'Mickey. Mate. Word gets around, when you've got the sort of previous you're wearing. I think you'll find *everyone* knows you're back in London. Now, what about we find you something to do, eh?'

4

The day after Kev's visit. Early evening, Monken Park High Street. The scene of Mickey's first kill. A street dealer who'd sold his sister the tablets that had killed her. Getting his methylenedioxymethamphetamine and his paramethoxyamphetamine mixed up. One of them being the rave drug Ecstasy. The other the slower acting and more intense Doctor Death. Its longer onset leading inexperienced users to take more. And thereby to OD. Overheat, collapse and occasionally die. Exactly what had happened to Katie.

His sister's death resulting in his murderous rampage through the OCG that had dealt the drugs. Revenge that had resulted in him leaving the Met, his marriage and the country. Mickey walked past the scene of his first kill. Flicking a single glance at the alley. Grateful to be done with all that. Just looking to put his feet back on the ground. And, without that lethal purpose, work out what to do with the rest of his life.

He walked into the pub to find the Friday Night Boys in attendance. All three of them once his beat buddies. Back in the day, when life had been so much simpler. All three of them now former coppers.

Den, still installing windows. Still apparently baffled by

just about everything that wasn't laid out in black and white. And still a little bit on the spectrum. Which was what had made him a decent street copper. If of little use for anything else in the Met's employ.

Steve, out of the Job but fallen on his feet. Having shopped his bent as a nine-bob-note gaffer DI Jason Felgate to Professional Standards. And subsequently found himself warmly encouraged to take early retirement. With beneficial financial terms and generous re-employment assistance. No reference too glowing for a man the force wanted to get shot of.

Turned out that too many mid-ranking coppers knew who'd turned Jason in. The term 'anonymous whistle-blower' their oxymoron of choice. And so Steve was on the street. Disgruntled but financially better off than if he'd done his thirty. And doing nicely. Managing security for a Merc dealership. Decent package, company car. Gleaming.

And Deano. Still the same old dinosaur he'd been as a street copper. Just no longer a copper of any variety. Still getting over the back injury that had medical discharged him. Pushed over a wall by a protester on a trans rights march. Part disgruntled, with two vertebrae surgically fused. Part grateful, what with the disability pay-off and early pension. And still completely baffled as to what all the LGBTQ+ fuss was about in the first place.

He was holding forth on the subject when Mickey slid through the door. Den giving Mickey a smile like he'd just come back from a weekend away. Steve taking one look and jumping to his feet. One part delighted to see his mate and former protégé. One part needing to go to the gents' in any case. And over the moon with an excuse not to listen to

Deano's stream of consciousness. Deano having just started a reprise on his theme of why JK Rowling was right. Even if her books were a steaming pile of shit.

Mickey greeted all three men with the appropriate masculine reserve. Hugs all round but all very manly. Recognised that yes, their glasses were almost empty. And yes, it was definitely his round. Got the beers in and took his place at the table. To find all three of them watching him intently. He nodded, knowing what was expected.

'Come on then, let's get it over with.'

'What, so you piss off for a year and then expect *not* to get the third degree on the subject?'

Deano. Playing it aggrieved.

'Deano. Mate. Give the poor bastard a break?'

Steve. The traditional peacemaker even where there was never a state of war. More like an incessant grumble of distant artillery with Deano. And sometimes not so distant.

'All I'm saying is he can't just—'

'I think you should give him a break. After all, his missus has—'

'Whoa! Den! A bit early?'

Deano, horrified at the breach in protocol. Den shrugged. Having neatly defused the underlying tension. Even if unintentionally. And driven by the spectrum's trailing edge rather than emotional intelligence. He gestured to Mickey. 'It isn't bothering Mickey. Why do we need to dance around—'

'Because, you muppet, he probably still misses her.' Deano's volume gradually creeping up into the sort of territory that turns heads.

'I do miss her.' Mickey knowing that speaking quietly would notch down the volume. And take their attention off

Den. Who was, after all, right in his summation. 'I miss her. But I realise she was right to leave me. Given what I put her through.'

Which got heads nodding. All three having been involved in the nuclear response by which Mickey rescued her. And knowing that Roz's abduction by Mickey's enemies had been the last straw. 'It's one of the reasons I was away for a while.'

'What, making up for lost time?'

Fresh eye rolling at Deano's leer. Deano being infamous for his pursuit of any woman not quick enough to detect and avoid him in his self-appointed role as 'The Monken Park Stud'. Brazen in both pursuit and trumpeting his use of erection-enhancing pharmaceuticals.

'No, Deano. Getting over losing her.'

'In Thailand? I fucking well bet you were! You were probably riding a different—'

'Deano. Shut the *fuck* up and let the man talk?' Steve. Amused, in an irritated way. Turning their attention back to Mickey. 'You're looking hench, son. Weights?'

'Training. I took up with a martial arts gym and did a fair bit of work.'

Which sparked Den's interest.

'Muay Thai? Looks nasty, all that kicking.'

Mickey nodded. 'Some of that. And some other stuff.'

'So now you're what? Bruce Lee's brother from another mother?'

Mickey shook his head patiently. Reckoning that when it came to not being self-aware, Deano gave Den a run for his money.

'No, Deano. I learned a few moves, sure. But all that Kung Fu stuff takes a lifetime to learn.'

A moment of silence. The conversational arc not ready to dip down into the usual inconsequentials.

Den to the rescue. 'So what are you going to do then? It's not like your skills won't be in demand.'

Mickey shrugged. 'Funny you should ask. I had Kev Smalls round last night asking the same question. He gave me an ad to look at.'

He fished a folded piece of printer paper out of a back pocket. Opened it out and put it on the table. LinkedIn logo at the top. Sterling Assurance Services. In no way riding the three-letter acronym gravy train of course. An 'exclusive close-protection agency'. With 'experienced and discreet close-protection personnel' sought for employment and long-term contracts. Proud that 'we provide a range of security and guarding services to some of the most exclusive clients in the UK'. Just call this number. Discretion guaranteed. And so on.

'Looks right up your street.'

Deano, unable to hide a trace of green-eye. Taken in by the apparent glamour of the job. Although to be fair, most Prot retirees were too.

'Yeah, probably is. If I want to get back into it.'

Mickey knowing his interview was already booked. Seeing the hook dangling before him. And knowing he'd probably have to take it.

5

James's first few weeks as an MI5 officer had been uneventful. Unsurprising, given he'd been recruited into a military liaison role. Working within H Branch. Corporate Affairs. Or, as James was starting to suspect, Corporate Arse-covering. His role being to make sure that military assets used to prosecute targets that were outside the Service's intelligence remit knew what to expect. How and where to deploy, obviously. And more importantly, the rules of engagement.

Critically, to ensure that those rules were obeyed. Or at least ideally. Worst case, to make sure that the rules were briefed. And that they *should* therefore have been followed. Documentation signed and stored. Responsibilities crystal clear. Backsides safely covered. And who knew better than James what a wise precaution that was?

After the brief interest of on-boarding, he'd endured a month and a half of mostly boredom. With nothing much to do. Other than keeping up with the general intelligence updates. For which he was clearly categorised as 'need to know (but not very much)'. Having already reintroduced himself to a less than interested Regiment.

But then it happened. Just as the shine on his new job was starting to dull. Unceremoniously shunted up two floors within Thames House. Without warning, first thing on a Saturday. Called into the office at no notice. Collected at reception, where he'd been instructed to wait on arrival. And pitched into a bear pit. At least by comparison with H Branch's hushed and cautious world.

G9C. Domestic Islamic terrorism a speciality. And a world away from Corporate Affairs. The operational department responsible for preventing the sort of thing that had happened the previous day. Five policemen killed on the M11. Incinerated by something far more devastating than simple explosive. Some of the better-informed speculating that the blast zone and damage done added up to something military. Something big and nasty. The police on the spot never – as the tabloids were screaming – having had a chance.

His escort walked him into an open-plan area containing thirty or so desks. All with double screens. All occupied. Serious intelligence horsepower. James was still wondering what on earth he could add to the organised chaos when he was ushered into an office. His guide depositing him at the door. Heading off into the hurly-burly with evident purpose. Having shrugged off the new guy.

'James Cavendish?' The woman at the desk didn't look up from her screen. Crisp white blouse, subtle make-up, expensive earrings. Hair cut in a low-maintenance bob. 'Susan Miles. I run this team. Forgive me for multi-tasking, I'm in a meeting with the minister, my boss and my boss's boss in Marsham Street in forty-five minutes. All three of whom are going to be looking at me in the expectation of answers. And these briefing slides aren't going to write themselves.'

Marsham Street. The Home Office. James nodded, taking a seat.

'You're wondering what the hell you're doing here, no doubt. I can probably help with that. Does the name Pavel Salagin mean anything to you?'

It did.

'Salagin's an oligarch. One of the originals. Became obscenely rich by palling up with Yeltsin, back in the nineties. And decided to relocate to London when Putin took power. Probably because he could see the writing on the wall. He owns Rossiya Racing, two yachts, five aircraft including both an Airbus A318 and a Spitfire, and several dozen very expensive cars. Total wealth somewhere north of ten billion. And that's about the limit of—'

'Good enough.' She shot him a terse smile. The message simple: *I don't have the time to waste.* 'Salagin's been under low-intensity surveillance for quite a while now. One of several suspects in the Winchester case.'

James nodded. Familiar with the events in Winchester the year before. A former KGB officer killed in a gruesome fashion. Murder with a 'Trotsky's ice pick' motif. Novichok having proven way too hit-and-miss. Found tied to a chair. With a log-splitting spike hammered deep into his head. And the murderer never found, although plentiful DNA had been recovered from the scene. The killer having taken the time to excrete on his corpse. DNA with a high percentage of Slavic gene markers.

An obscenity sending two remarkably clear messages. If the reality of the Skripal attack hadn't sunk in yet. One: open season on former Soviet traitors had most definitely been declared. Two: it takes more than one person to kill a man like that. More like a team.

'We suspect him of harbouring the murderers?'

Her mouth twitched in the suspicion of a smirk.

'*We*. Nice touch.'

James nodded, straight-faced.

'I was an army officer. It's—'

'All about esprit de corps? I always thought it was about group-think.' She grinned, mirthlessly. 'I'm fucking with you. And while we do indeed suspect him of having assisted the murderers, we've now got something a lot more serious than that to worry about.'

The penny, already teetering, dropped.

'This is about the M11 deaths.'

'The M11 *murders*. Yes, it is. They might have been a nasty shock, but they weren't entirely unexpected. We've known something like this was coming for weeks. Not what, of course, or we'd have set the dogs on them. There's been chatter. Electronic activity by suspects, indicating that cells across the north of England were being woken up. As if something or someone was enthusing them, one at a time. Which in turn started generating the classic message patterns, one cell at a time. Gathering the faithful to their holy duty.'

James passed over the use of 'cell' as too complex to unpick in a few minutes. A word implying something more than simple home-grown terrorism. Something worse than the amateur and inefficient young second-generation jihadis of yesteryear. The young men whose bombings had galvanised the Service in 2005.

'I thought they all used coded message platforms these days?'

'Oh they do. But even if we can't read the messages, we can see them being sent. And we had surveillance teams on the most likely suspects. Including the M11 bombers. Waiting for them to gather and give us a reason to arrest them all. What blindsided us was the means by which they received the weapon. We were holding off making any sudden moves, because there were no signs of any weapons or explosives on their premises.' The lightness of her phrase made it clear she expected him to know that burglary was a routine Security Service tool. 'What we weren't expecting was for a third party to turn up with something rather special for them.'

'Explosives.'

'Exactly. Except this wasn't just enough plastic to peel a bus open, horrific though that would be. This was a Russian Sunburn warhead. Enough to vaporise the entire bus.'

James knew all about the Sunburn. Its Russian name, *Solntsepyok*, meaning 'blazing sun'. An ironic name in itself. Military black humour in naming an infantryman's worst nightmare. Napalm for the new millennium. A rocket designed to be fired by the dozen from a box launcher on a converted T72 tank. Three such tanks sufficient to kill an entrenched infantry company to the last man. The most robust of field fortifications no defence against their incandescent heat and monstrous overpressure. Already field-tested in Chechnya, according to the intelligence reports. And then used in action in half a dozen wars. He nodded his understanding.

'And if my team look quite motivated, it's because it wasn't only police officers who died yesterday. One of our own didn't go home last night either.'

'Dead?'

She nodded. Dry-eyed, hard-faced.

'He stopped a piece of the warhead's casing. Right between

the eyes. He was dead before he hit the ground. So if that lot out here seem a little touchy, you'll have to—'

'Make allowances.'

He met her stare levelly. Sending a message back. *If interrupting's good for you, it's good for me. And rank be damned.* She nodded, apparently untroubled. A test passed, perhaps.

'Yes.'

'So when I'm done making allowances, where do I fit in here?'

She nodded again. Perhaps approving of the no-nonsense down-to-business approach.

'Like I said. Salagin looks a lot like someone who could shelter whoever it is that's driving round the country handing out military ordnance. He certainly has the resources. And we think he's a lot closer to the Putin regime than he lets on. So we want to put eyes on him, close up.'

He shook his head. Slightly baffled. 'But where—'

'Do you come in?'

James nodded, resisting the urge to ask her to stop finishing his sentences. Knowing that she'd probably just accuse him of whatever man-verb combination was fashionable that week. Man-whingeing most probably.

'Don't worry. The person I want to put in to keep an eye on him isn't you, Major Cavendish.'

6

Mickey found himself sitting in an armchair. Having been escorted to the large, airy sitting room by a fresh-faced young woman dressed in formal business wear. Low heels hinting at a sexist-bullshit-free environment.

'If you'd like to wait here please, Mr Bale? Mr Salagin will be along to see you shortly.'

Mickey smiled and nodded agreement. Waiting to take a seat until the assistant had left the room. Basic manners, he supposed. Tempted to walk around. Examine the various expensive-looking items decorating the shelves. Or to stroll over to the window. Watch the comings and goings in the street. Rollers, Bentleys and the top-end German limos ten a penny in this neck of the woods. And resisted both temptations.

Knowing for one thing that he didn't want to be caught eyeballing a Fabergé egg. Or be taken for a man without the patience to simply sit and wait. And that the odds he was on camera were better than good. And so he waited. The spirit of good-natured patience. Or at the least a man who recognised a test of his patience when he saw one.

It had all been Kev's idea, of course.

'You sit round here much longer, you'll turn into a fucking potato. Get yourself out there! Make some money, get some purpose. See some fucking life, eh? Have a look at this!'

Mickey had been reading the same ads for most of the last fifteen years. Initially envying anyone who could qualify for the lifestyle portrayed in their gushing prose. Then, bit by bit, he had revised his opinion. As his experience grew. And as he listened to colleagues who'd actually gone private sector. None of them ever describing it as the land of milk and honey. Life after Protection Command not exactly the gravy train of Prot officer fantasy. Not faced with the cold reality of the private sector.

Serving coppers inevitably declared themselves bored with routine. Quacked on about working for thankless politicians. Mistakenly allowed themselves flights of fantasy as to how it could be. Big houses, flash cars, yachts, perpetual sunshine and money aplenty. And women, of course. Glamorous women. All probably eager for a bit of rough on the side.

Whereas what they tended to end up with was just equally thankless private clients. And women who wouldn't give them a second glance. Most ex-coppers being somewhat nondescript compared to their glossy principals. Who also had the undying attraction of wealth. While most Prot escapees were comparatively skint.

The private sector tended to feature hours that would have Prot rank and file up in arms. And boredom. Soulless, grinding, crushing boredom. At least Prot work was varied. And usually interesting, off and on. And at least with the politicians, what the protection officer said was law. The new boys and girls usually too inexperienced to know they could have a different opinion.

Which meant that politicians tended to be housebroken. And to do as they were told. Unlike the royals, whose life experience made them more familiar with how it all worked. Which rules could be stretched, bent, and on occasion broken.

Somewhat different, in the private sector. Where principals tended not to listen to perfectly well-founded advice very closely. Or at all. The word 'instruction' never being used if a bodyguard knew which side their bread was buttered. And that was just the principals. Who, Mickey had come to understand, were the easier half of the equation.

Wives, girlfriends and older children being by far the harder part of the job. At least the principal sometimes recognised at least the kernel of his bodyguard's role. Whereas their dependents frequently utterly failed to either know or care about it. Too wealthy. Too privileged. Or, occasionally, just too stupid or entitled.

Blind to what might happen if the wrong people got hold of them. Too concerned with TikTok to give any thought to what to do if it all went BangBang. Which was fine with most of Mickey's former colleagues. Who had, for the most part, swiftly become hardened to such a possibility. As long as it didn't happen on their watch.

Which was why Mickey had gone to see the agency in a cynical frame of mind. An address in the heart of Mayfair. Unfashionably tucked away, but crisp and well presented. There being nothing down at heel about Sterling Assurance Services. And found himself on the other side of an immaculately clear desk. Sitting opposite a man who could have been James Cavendish's older cousin.

Roderick Smythe, former Intelligence Corps. And current provider of 'the highest quality protection officers to the most exclusive clients in the capital'. Who declared himself more than a little pleased to make Mickey's acquaintance.

'I'm absolutely *delighted* to meet you, Michael!'

Mickey read the runes in the use of his first name. Sat and listened with the appropriate show of interest. As the immaculately tailored ex-officer explained that Mickey was exactly what he'd been looking for all week. Hence, Mickey guessed, the speed with which he'd been invited to come for coffee. High-quality protection officers being something of a rarity in the guarding market, it seemed. Which was, Smythe informed him, overwhelmed by a flood of former coppers and soldiers. All drawn to what looked like easy money, from the outside. A precious few of whom, Smythe explained, really were quite good. Although much of the remainder weren't. All somewhere between mediocre and hopelessly unsuited for anything other than door work.

'And there's nothing wrong with door work, of course. It's just that we simply couldn't afford the blowback if one of our chaps punched out a pap, for example. You've got a decade of experience in defusing that sort of situation without excitement. Which is why I'd be very happy to put you on our rather exclusive payroll.'

Mickey had listened to enough embittered employees of guarding companies in his time. Which meant he knew how permanent employment worked in the industry. And had therefore politely declined the offer.

'I think my days as a salaried employee are behind me, thank you. And given this setup...' he'd waved a hand at the office, as elegant as it was spacious, '...you'll be charging my services to the client at what, fifteen hundred a day?' He'd fixed the older man with a direct stare. 'After all, it's not like the rate will stay secret very long. It never does. And the minute I feel like I'm being exploited, I tend to vote with my feet. So I think an open-book contract agreement would be more in keeping with the potentially temporary nature of any work I do for you. A zero-notice contract.'

Smythe raised an eyebrow. And then laughed. Recognising that his bluff had been called.

'As you say, it never stays secret. And fifteen hundred a day? Not bloody likely! You're a George Cross holder, man! I think I can winkle two and a half grand a day out of this particular client. If, of course, your face fits. So, shall we say a thousand a day? On twenty-eight-day terms?'

Which had made it Mickey's turn to raise an eyebrow. And Smythe had made the mistake of betraying his eagerness by being the first to break the silence.

'Very well, a straight split? Fifty-fifty? Can I say fairer than that? After all, as you point out, premises like these do cost…'

Mickey looked pointedly at the Patek Philippe Nautilus on the colonel's wrist. Then checked the time on his rather more prosaic Tudor Black Bay. Albeit a limited edition. The one with the Royal Household's crest painted on the dial. A sure-fire door-opener, if such a thing could be said to exist in the business. Sending two messages.

One being that Smythe would be wise to stick to Rolex, inside work hours. Keep the Patek for his club and regimental reunions. The other message being a swift reinforcement of his impeccable credentials.

'I'd prefer a sixty-forty split. My sixty per cent billed to you weekly, and paid no later than seven days after invoice. With the clear understanding that no pay means no show.'

Smythe had smiled wanly and put a hand out.

'Agreed. You'll be wanting a written contract?'

Mickey shook his head. Taking the hand and sealing the deal.

'No need. I think you'll pay up dead on time, if this client is as important as you say.'

All of which was the brief precursor to him walking a few hundred metres to the next stage of the interview. Upper Grosvenor Street. The London residence of one Pavel Salagin. A name known to Mickey from his interest in motor racing. Rossiya Racing being one of several mid-table teams vying for best of the rest.

'Mr Bale.'

Mickey looked up to find the man himself standing in the doorway. Wearing a smile of welcome. A youthful face for his age. Sandy hair with a scatter of grey. Blue ice chips for eyes. The moment at which many people would find themselves a little overwhelmed, he guessed. Buckling in the presence of what was undeniably greatness, of a sort. Not Mickey. Having bodyguarded enough very well-known people to give no shits for fame. And little more for fortune, other than respect where due. He stood, taking the offered hand. Meeting the friendly appraisal head-on.

'Mr Salagin.'

'Please, sit.' The Russian's accent was so light as to be almost unnoticeable. He took a seat opposite Mickey, shooting a swift glance at his watch. A Richard Mille. The usual open-faced and complication-strewn dial. Its hands almost invisible against the intricate backdrop of wheels and springs. He saw Mickey looking at the watch and smiled again. 'Do you know your watches, Mr Bale?'

Mickey nodded. 'Enough to know that what's on your

wrist is probably worth a quarter of a million pounds, Mr Salagin.'

The other man grinned, leaning forward and pointing at the Black Bay. Mickey's watch half hidden by the cuff of his shirt. The royal household's rose crest barely visible.

'So would it surprise you to know that I would exchange it for that Tudor you're wearing, here and now? I would take it off and offer it to you now, if it wouldn't immediately make me seem like a vulgar arriviste.'

Mickey, with enough education to guess what he meant, nodded his understanding.

'I couldn't part with it in any case.'

Wondering what the fuck he was spouting even as he said it. But the Russian bowed his head in acceptance. His continued smile telling Mickey that he might just have passed some sort of test.

'Of course not.' His voice was as soft as before, but invested with an emotion that Mickey was slightly surprised to realise sounded a lot like respect. 'It has a sentimental value to you. It reminds you of your service to the state. And your exploits of not so long ago, yes?' Salagin having clearly done his homework well. 'So, tell me, Mr Bale. Are you a Michael or a Mickey?'

And perceptive too. Knowing his British social mores all too well. The outsider looking in, perhaps?

'On duty I am always Michael, Mr Salagin.'

Not invited to use the potential client's first name, Mickey had no intention of making any such schoolboy error.

'And off duty you are Mickey, I presume? I ask because the dossier that I have on you states your preference for the less formal version.' The Russian smiled, unabashed by Mickey's level stare. 'Forgive me. It was a slightly heavy-handed way of making it clear to you that I tend to be very well informed on all matters that are important to me. I subscribe to various intelligence databases. One of which provided me with a helpful summary of your career with the Metropolitan Police Service's Protection Command.' Mickey kept his face perfectly composed. Waiting to find out if the database in question was as well informed about his extra-curricular activities as Kev had been. 'A career brought to its conclusion by your being wounded in defence of your principal a few months ago. And resulting in the award of this country's most prestigious non-combat medal.'

Mickey nodded. 'Even if it felt a lot like combat at the time.'

Salagin smiled. 'I'm sure it did. And you carry its mark with as much pride as you do the medal, I suspect.' He looked at Mickey for a moment. 'So why is it that such a highly decorated policeman leaves the service he so clearly loved? If my source is to believed.'

'Don't believe everything you read, Mr Salagin. The Metropolitan Police and I had come to the end of our association.'

The Russian nodded. Then gestured to the room around them. 'And *this*? Why do you wish to serve as a protector of a man not born here?'

Mickey shrugged. 'Apart from the money? I protect people, Mr Salagin. It's what I do. Pretty much all I can do, truth be told. And your origins aren't my concern. Just your safety. If you decide to use my services.'

'I see.' Salagin leaned back in his seat. His eyes hardening, as he regarded Mickey for several seconds. 'Let us be clear, Mr Bale. In you I see a missing piece of the jigsaw that is my public profile. I am watched all the time, as you can imagine. When I attend a motor race at Silverstone. When I shop for a new watch in Bond Street. When I go to Ascot for a day's racing. When I do all of these things, I am on very public display. Photographers take pictures of me, to put in the newspapers. The rich and aristocratic men at the top of your society look through their privilege-tinted glasses at me. And everyone else shares some aspect of that judgement. There's Salagin, they tell each other. A Russian nouveau riche bastard and English gentleman manqué. Do you know this word, *manqué*?'

Mickey shook his head, and the Russian raised a hand in apparent apology.

'I ask the question not to show off, but simply to demonstrate the lengths I go to in order to be accepted. Half the people who look down at me have no clue what it means either. It means to be frustrated in an ambition, Mr Bale. And, were they to call me an English gentleman manqué, they would be correct. If a little unsympathetic. I have lived here for twenty-five years now. I made a lot of money when I was a young man. A *lot*. And I chose London to be my adopted city because of its receptiveness. Expecting it take me to its heart as a valuable new citizen. Only to realise somewhat later that your country is rather more receptive to money than to the immigrant bearing it. Whether rich or poor. And so I go out of my way to demonstrate that I am as English as the rest of you. That I speak your language better than many of you. Have a better house than most of you...' He paused, smiling, and when he resumed his voice was as level as it had been when he'd greeted Mickey from the doorway. 'So, Mr Bale, do you discern my purpose?'

'Having a retired Protection Command officer at your side makes for good PR?'

Salagin nodded. 'Exactly. And yet at the same time you have proven skills that may one day come between an attacker and me. It is, as the Americans have taught us to say, a win-win. When I attend a black-tie dinner I will expect you to escort me, with your medal in place. In doing so you will add the cachet of your achievements to my reputation. My enemies will call it vulgar of course.' He shrugged, evidently untroubled. 'But it will serve my purpose admirably.'

He stood, pulling his cuff back to look at the Richard Mille again.

'And now I must leave you to my staff, I am afraid. My bankers have a tendency to like punctuality, like all of their kind. You will join my household, I presume?'

Mickey nodded. 'Providing you can come to an agreement with Sterling Assurance Services.'

Salagin smiled. 'Colonel Smythe and I already have an agreement, Mr Bale. He overcharges me by five hundred pounds a day, and I let him do so on the understanding that I always come first among all of his clients.' He headed for the door. 'I will ask Ms Roberts to introduce you to my head of security. Be prepared for him not to be quite as well naturalised as I am.'

7

L ate lunchtime. A West End bar. Mickey having ordered
the drink he knew his guest would choose were she there.
Classic margarita. Sipping at a Peroni. Keen to keep a clear
head for the afternoon.

'Well now, Michael Bale.'

He turned to greet her. Finding her much the same as the
last time he'd seen her. Finely drawn features still lightly
etched with the grief of loss. Her dark hair cut short, clothing
practical rather than stylish. Better suited to crawling
through bushes with a long lens than drinking cocktails in
Knightsbridge.

'Not all that far from the last place we met. But it looks
like you came a long way.'

Mickey pirouetting on the spot. Knowing that the new
Mickey was both trimmer and hencher.

'I went to Thailand for a month. Turned into a year.'

The barman delivered up the cocktail. Which was
tasted and declared more than satisfactory. Correct quality
ingredients. Correctly mixed. Tamara being pretty fussy
about her margaritas.

'So why come back?'

A blunt question. Bluntly asked. Tamara having no special emotional link to Mickey. Indeed some part of her still holding him responsible for a close colleague's death. Not vindictively. Just factually. The fact that he'd saved her from a similar fate or worse a factor too.

'Felt like I had unfinished business. And besides... Thailand...'

'Pleasant enough, but just not home?'

He nodded. Why else come back to Monken Park? And not somewhere a bit nicer?

'Yeah. You been well?'

'Pretty good. I'm not at the *Clarion* anymore though. Anthea made the mistake of saying how people vanish in London all the time one time too many.'

'Ah.'

Mickey had wondered whether Tamara could restrain herself where the news editor was concerned. The promise to break her nose if she said the wrong thing apparently not bravado.

'So did you catch her properly?'

'Oh yes. I felt her nose break. She threatened to "caw da bolice". I threatened to release the video of her colluding with that policewoman.'

Mickey's ace in the hole. And all that had stood between him and prosecution a year before. Video that proved his assistant commissioner had leaked information – well, lies – to the press.

'A lucky day, the day you shot that footage.' She looked at him for a moment. Mickey cursing himself. It having been the same day that her colleague Mark was taken. Kidnapped by an angry gang boss. Tortured and murdered.

'Sorry. Crass of me. So what are you doing, if it's not for the *Clarion?*'

'Same same.' She sipped at the cocktail. 'Pap work, mostly. Ultra-wide in your face or long lens from half a mile away. Anyone famous, or better still infamous.'

'Pays well?'

She shrugged. 'It's what you make of it. A decent street shot of a reality star out shopping can fetch two hundred and fifty quid, if they use it. The same celeb caught falling into a taxi at two in the morning pissed, perhaps two grand. Add in a member of the opposite sex they're not married to, make that five. If the other party is already partnered, make that ten. And if they're getting jiggy in the back, it could go as high as fifty.'

'Hard work?'

She shrugged. 'Not if you're hungry enough. Or just have a mortgage to pay. Wouldn't want to be doing it at fifty, right enough, but it's fine for now. And who knows, one good pic could pay off my mortgage.'

'So where does Pavel Salagin figure on your targeting list?'

'Salagin? The F1 boss? Why do you ask?'

Mickey shrugged. Keen not to show too much of his hand, even to a friend.

'I was just reading about him.' And left it at that.

'Salagin.' She thought for a moment. 'Not much interest, really. Not unless it all gets exciting with his racing team. He doesn't frequent night clubs; he's not a substance abuser from what's known. Seems to take his pleasures "in private", so to speak. It'd have to be a pretty special pic. You sure there's no other reason for asking? You're not—'

'No.' Mickey put his best blank face on. 'Private close

protection is for losers. Just imagine a horse piss cup-a-soup with dogshit croutons and that's pretty much it. And it's not like I need to work for a living. Not with my simple tastes.'

She looked at him over the glass's rim.

'So, what's your plan for the rest of the day?'

Mickey knowing he wasn't being propositioned. Not Tamara's style, or within the bounds of their relationship.

'Lunch with you, obviously. I've booked a table. After that I thought I'd go clothes shopping. Nothing special, just a bit of a mooch up Oxford Street.'

8

Mickey let himself back into the flat that night with the usual sinking feeling. Dropped his shopping bags in the hall. Having hung around in town for a while specifically to avoid the place. Bought some clothes for his new job. Made a drink and dinner last until the inevitable could no longer be fended off. And gone home to his new abode.

One bedroom, a bathroom, lounge and kitchen. Little more than a place to sleep. Marooned on an intermediate floor. At the mercy of other people's overflowing bath water and noise pollution. No view other than the block next door. Not very much better than the old police section house he'd started his career in. Newer build, shoddier built. He took the key from the lock and closed the door. Put his hand to the light switch.

And froze.

Standing in the darkness with the vaguest hint of illumination. Street light from the lounge window through the half-centimetre crack between door and frame. Complete quiet, other than the faint sound of conversation from the flat above.

Something had his hackles up. No sound. No aroma. No light. Something else. He stood in the small lobby, waiting and listening. Wondering what had put him on alert, when there was demonstrably nothing—

A tiny sound. The minute creak of leather. A sound he heard every time he shifted position in the old chair facing the TV. Options presented themselves. He could investigate. Be ready to fight. Or he could go back out of the door. And make a run for it. Probably the best option, unarmed and in the dark. Because blades. Because silenced pistols and night-vision gear. And because the list of people with an interest in his unwell-being wasn't a short one.

For one thing there was a foreign government whose operators had been killed to save his life in Belize. An implacable enemy, if unlikely. Then there were the comrades of the assassins he'd killed in the Fulham Road a year before. Violent men who might well come seeking vengeance. Or the remnant of Castagna's gang, perhaps, looking for exemplary revenge. Probably better not to find out which it was.

The decision needed no conscious thought. Get out. Dial three nines. Let the police sort it out. He turned stealthily back to the door.

'Are you going to stand there in the dark *all* night, Michael?'

No *fucking* way. He flipped the light switch and walked through into the lounge. Lit that up too. Revealing what he'd known from the intruder's first word.

'James Cavendish.'

The other man looked up at him. Staying seated, hands on the armchair's wings. Deliberately non-threatening.

'Good evening.'

Mickey shook his head.

'I nearly called the police on you. Want a cup of tea?'

Cavendish smiled. The same boyish grin that had deceived Mickey as to his capabilities, at first.

'Kind of you, yes please. The police wouldn't have come, of course. I've put a 999 block on both this address and your name.'

Mickey stopped in the kitchen door. Making connections. Realising what he ought to have known months before.

'You put a block… Jesus. You're *Box*?'

The seated former officer nodded equably at the question. Smiling faintly at the archaic slang term for MI5, still beloved of both the armed forces and police. 'Box'. For PO Box 500, the Service's wartime address eighty years before. He chose to use a somewhat more formal term in reply.

'Right in one. I am indeed an officer in the Security Service.'

Mickey put the kettle on and turned back to the lounge. Checking that Cavendish was still seated, for one thing. Making connections. And wanting to see the other man's face, and gauge his truth, for another.

'You always were, weren't you? When did you *really* leave the army?'

Which got him a terse laugh.

'That's need-to-know information, Michael. And you don't have any such need.'

Mickey shrugged. 'Explains a lot of things though. Doesn't it?'

Cavendish smiled. 'Count yourself lucky. Most people who work with men like me never actually know who they're really dealing with.'

'Lucky me.'

Ignoring the other man's sardonically raised eyebrow,

Mickey busied himself making the tea. Carried two cups into the lounge, passing one to Cavendish.

'Here. I promise the spoon's been nowhere unnatural.'

'Thank you. And it's good to see you looking so well.'

'Given I was gut shot the last time we parted company?' Mickey nodded. 'I should have realised there was something iffy about you. Especially when I realised you weren't even in the same hospital as me and Kev.'

The other man nodded. 'Indeed. I was whipped away in the middle of the night, once the rather intrusive reconstructive surgery on my shoulder was done. Private clinic, high security. And not just for my own safety either. For a while there were strong suspicions that I might have been the source of the leak as to the convoy's route that night.'

'What, after what you did? I'd be dead if not for you. And so would the Home Secretary.'

'And you're very welcome, Michael. I did rather lean on that fact when the interrogators came to call.'

'But you're cleared now?'

'So it would seem. Five years with the Regiment probably helped as well. And was enough to get me recruited by the Service. Initially into a low-profile role. Making sure that my former colleagues toe the line when operating under our authority.'

Mickey stared at him for a moment, putting two and two together in a way that felt painfully slow.

'Initially. But now you have something more *high*-profile, right? And, given the time of night, that's probably what you're here for. Isn't it?'

Cavendish nodded. 'Yes, Michael, it is. Any my apologies.'

Mickey leaned back, sipped his tea and grimaced at the

temperature. 'Let me guess. The agency that put me up for the Salagin job is MI5-controlled?'

The other man shook his head with a knowing look. 'Oh hardly. That would stick out like a sore thumb. We find Roddy Smythe's combination of a suitably thick upper crust and palpably naked greed a much better cover. The agency is genuine enough… just supportive of our aims.'

'But someone had to apply the nudge that put me in touch with them. Which means Kev, right?'

'Mr Smalls was recruited by the Service, once he was fit enough to ride a motorcycle again. We use him for the lower-profile jobs. Moving sensitive items securely at speed. The occasional vehicle-follow job, when the target isn't expected to indulge in any counter-observational tricks. As a former police officer we know he can be trusted. And he can give some of our younger riders a run for their money, when it comes to London games.'

'And he does the occasional bit of recruitment too?'

Cavendish had the grace to look uncomfortable. 'Not that he needed very much encouragement. I think he believes you're somewhat lacking in impetus.'

Mickey raised a jaundiced eyebrow. 'I need the sort of *impetus* you have to offer like I need a nine-millimetre enema.'

The other man leaned forward. 'Mr Smalls doesn't actually do recruitment, as such. He simply pointed you in the right direction. Encouraged you to go and get the job he wanted you to have. And you went and got it all on your own. It's now *my* job to recruit you into MI5.'

After a long moment of disbelieving silence Mickey shook his head. 'And if I don't choose to be recruited? I mean, no disrespect intended, but I've already taken one bullet for you lot. Not to mention coming within a second or two of having my throat cut.'

Cavendish smiled. 'You can choose not to perform this task for your country, Michael. I have to warn you, however, that your country will be somewhat disappointed. And very obviously so.'

Mickey stared at him in silence. Waiting for him to continue. Got bored with the silence disappointingly quickly. 'That was a threat? I didn't just imagine it?'

'I'm afraid so.' The other man raised apologetic 'what can you do?' eyebrows. 'It does rather go with the turf, I'm afraid. It's quite rare these days for anyone to volunteer to do little jobs on the side for us. It seems there's a good deal less patriotism than there used to be.'

'That, and a good deal more awareness of the fucking horrible consequences of being caught doing those little jobs. Because being deniable tends to result in being denied. Fast-forwarded straight to accelerated interrogation and prison. Hard regime. Without even the expectation of being swapped for one of theirs. Because you're not one of ours.'

'That too.'

Mickey crossed his arms and waited. 'So...'

'You want to know what the threat is?'

'Well it wouldn't be much of a threat if I didn't know the details. Would it?'

Cavendish nodded. Taking a phone from an inside jacket pocket. 'Commendably straightforward of you, Michael. But then you were always quick getting to the point. So yes, to business. It seems that you left the Metropolitan Police Service under something of a cloud. Suspected of a quite exceptional series of violent revenge crimes. Quite a killing spree, in fact. Whose punishment, were sufficient evidence ever to come to light, would be enough to jail you for decades. Apparently one Deputy Assistant Commissioner Haskins of Professional Standards was already committed to charging you. Speculatively, it seems, but with high hopes of achieving a successful prosecution.'

Mickey nodded. 'I wondered where Kev had got that little snippet of information.'

'And then, just as the DAC was getting ready to intone the last rites on both your career and your freedom, he was frustrated. Your assistant commissioner intervened. Case closed, no further action. Sergeant Michael Bale to be medically discharged as physically unfit for duty. The lack of any solid evidence being enough to ensure that your retirement was officially deemed honourable. And with a George Cross to sweeten the pill.'

'What can I say? Their piss-poor collection of half-baked evidence was—'

Cavendish turned his phone on, then showed the screen to Mickey. Revealing an image of several heavily armed figures. Clustered around an exterior door, at night. The night of Joe Castagna's death.

'Does this scene perhaps look familiar to you?'

Mickey, recognising the moment in question, resolved to shut his mouth and keep it that way. Wondering where the shot had come from. Slightly perplexed, given they'd trashed the hard drive before leaving the scene.

James gave him a tight smile. 'Going quiet won't help. We tracked this team down. It wasn't hard to do with the resources available to us. Found out that they were already known. Contingent contractors who have performed specialised tasks for our sister service. And then we worked out just *whose* weapons they used to do the job. Which, it has to be said, was an act of audacity verging on the foolhardy. But which meant that I was able to have a very informative chat with Mr Shaw.'

Naming the assault team's leader, with whom Mickey had cut the deal that had resulted in their taking down a gangster's operation. And ending Mickey's hopes that he might be able to bluff his way free of MI5's clutches.

'And Mr Shaw, faced with the government in question finding out he'd been in their rainy-day cache of firearms, was swiftly forthcoming. He even volunteered to surrender your share of the funds to us. A startlingly large amount of money, which explains why you were in no hurry to find employment. But I decided that it would be far more helpful in any prosecution of you all for your financial relationship to remain unchanged. I simply took all the information he had to offer and put it on file. A file under your name, with the classification of "agent leverage". So I'd say your choice is really very simple.'

'Co-operate or go to prison? I'd take Assistant Commissioner Chen with me.'

Cavendish gave a little shrug.

'I can't say I'd be all that troubled if you did, Michael. I'd imagine that whatever hold you have on Ms Chen paints her in something of an unflattering light. Oh, and that friend of yours? The one who left the country in a hurry? And who now lives in a Gulf State that currently lacks an extradition treaty with the UK? Covert rendition isn't just for terrorists, you know.'

Mickey shook his head. Knowing when he was beaten. 'So what is it that you want from me?'

'Nothing much. Come to my office, tomorrow. Watch a video that I'd like you to see. It's a little upsetting, but will illustrate the nature of the problem we have.'

'A problem that involves my new employer?'

'Quite possibly. And that, as you will have guessed, is the point of all this. We need a man on the inside. A man who is blameless, and of unimpeachable character. *You.*'

9

Late morning. The streets empty, for the most part. Those who were coming in to work staying at their desks. With delivery food to reduce their exposure to the latest variant. Salagin got out of the Mercedes onto a quiet street south of the river. Casting a wry smile at the quarter-million-pound car. A mobile extension of his gilded cage. The door held open by his security chief.

Yuri Kuzmich. A heavyset man in his late thirties. Bulky, even, but with the seemingly lazy yet lightning-fast reflexes of a big cat. Bearded, swarthy, scarred and dark-eyed. Dangerous.

'Pavel.'

He turned to look at the big man. Speaking Russian, as they always did when alone. Having known each other for the best part of three decades, both men knowing the other's likely thoughts and reactions.

'Yuri?'

'I will wait here with the car.'

The oligarch smiled. 'I doubt that this will take very long. It will be a meeting without vodka, I expect. And be kind to the traffic wardens. They're only doing their jobs.'

He walked into the building without looking back.

Knowing that the other man's eyes would follow him every step of the way between car and reception. And that as he passed out of Kuzmich's field of view he was entering alternative surveillance. Several uniformed guards stationed across the sprawling atrium. All with body cameras. Pinprick blue lights on their lapels and live streaming to who knew where. One of the receptionists stood up and came around her podium desk, a key badge in one hand.

'Mr Salagin. You're expected, sir.'

He took the plastic rectangle and clipped it to his lapel. Murmured his thanks and made his way to the elevator bank. Knowing to choose the one on the far right. Direct to the top floors. Exclusive to GazNeft. Not completely occupied, of course. But what company doesn't like to have a little growing space? And whatever went on behind the blanked-off sections of the upper office was nobody else's affair. The rumours said it was a clandestine trading operation. Rumours that the company had never sought to deny.

The elevator door opened as he approached it. Either proximity-sensing the card or under simple good old-fashioned remote human control. He stepped in, waiting impassively as the doors closed and the car whisked skywards. Knowing what to expect when the doors opened at the top of the building. The same carefully composed politeness that he got everywhere. Just not quite as sincerely meant as was usually the case.

Another podium desk, this one manned by a uniformed security operative. Just like the men in the atrium. But, unlike them, not British. Security on the GazNeft floors was looked after by former soldiers. Veterans of Russia's undeclared wars. All speaking decent English, all ex-Spetsnaz. All killer, no filler. He looked over the desk at Salagin. With no expression in eyes that had, the oligarch mused, seen death and horror at close quarters. And making who knew what of the man before him.

'Pavel Ivanovich. Welcome to GazNeft.' And there it was. The use of his patronymic immediate proof that he was effectively back on Russian soil. Among his fellow countrymen. 'If you'll come this way?'

He followed the uniform across the open working area, through several dozen appropriately spaced workstations. Arriving at a row of glass-fronted desks where a trim woman in her thirties was waiting for him. Her crisp white blouse and blue skirt not quite a uniform, but not far from it either.

'Pavel Ivanovich, welcome home.'

Salagin smiled. Unable to deny to himself that the words sounded good. Even knowing that was the intention. Even knowing that neither security man nor personal assistant considered him truly Russian. Not after so long away from the *Rodina*. He shot his immaculately tailored Savile Row worsted cuff. Taking an ostentatious look at the wafer-thin Jaeger-LeCoultre. He was barely a minute early. Looked over the personal assistant's shoulder and into Chasovshchik's office. Seeing the man he had come to meet sitting with his back to the door. Telephone to his ear.

'Thank you. Is he likely to be free any time soon?'

The gentle taunt spoken with a smile. The expression of a man who knew better than to complain. Making the point, nonetheless.

'He asks your forgiveness, Pavel Ivanovich. He has been unavoidably detained on a call. Senior management…'

She shrugged, a gesture of helplessness. And Salagin

nodded his understanding. Ah. That old euphemism. There was nobody senior to Sergei Ilyich Chasovshchik in the UK GazNeft office. There were certainly more important business roles. But Chasovshchik was at the top of the most important department in the building. Alternate Strategies. Almost comedic in its oh so obvious double meaning. Thumbing its nose at the British Security Service. *We know that you know that we know that you know that we're not just here to sell gas.* And not really caring.

Not caring that Russia's economy was only two-thirds the size of the UK's. Not caring that its people generated a quarter as much wealth per head as the British. Because Russia was free of the petty restrictions that the US imposed on her allies. Whether they liked it or not.

And Russia, free to act, had been re-energised by new leadership to do just that. And was waging war by other means with all the strength and guile it could muster. Hamstrung to a degree by sanctions. And hopelessly entangled in the corruption and drift of dictatorship. But still a dangerous foe, lacking any morals or restraint. Especially as only one side in the struggle actually realised it was at war.

And in Chasovshchik, Russia had a most diligent and effective servant. His nickname 'the Watchmaker' the literal meaning of his surname. And a reasonable approximation to his operational skill and patience.

'Pavel Ivanovich!'

Salagin stirred himself from the chain of thought that had been troubling him lately. Turned back to the office door. Chasovshchik waiting for him, hand stretched out. And, like the good Russian he knew had to be, he reached out and took it. Allowed himself to be guided to a chair at the meeting table. Chasovshchik sitting beside him. So close their knees almost touched. And in front of them, a slim box the size of an A3 envelope and as deep as a cereal box.

'Forgive me, Pavel Ivanovich. When my boss calls he tends not to be all that bothered about who he keeps waiting.'

Salagin inclined his head gravely.

'Matters of state will always take precedence.'

'Quite so. And it is good that you understand that.'

The other man fell silent, and Salagin recognised the signs immediately. That the other man planned to play the 'who breaks the silence first' game he loved so much. Having come to believe it made him seem even more powerful. Although it was in reality simply tedious beyond Salagin's patience. And so he conceded defeat so fast that it felt like a victory.

'And now, Sergei Ilyich, what can I do for you? I know what a busy man you are, so I won't waste your time on small talk.'

Or give you the opportunity to mention my family in St Petersburg for the fiftieth time, for that matter, he didn't add. The leverage that his two sisters and their children gave the other man. And the men who stood behind him.

'Straight to business, eh Pavel?'

Patronymic dropped.

'Straight to business.' A beat to let the other man know that he was up to the game. 'Sergei.'

Although I'd rather call you shithead, he mused inwardly. *So what the fuck is it that you—*

'You will be aware of the outcome of the *Prizrakis'* first operation.'

'Yes.'

Prizraki. The ghosts. And yes, he was. All too well aware.

Five men dead. Five families mutilated. Far fewer than he'd feared. More than he'd hoped.

Chasovshchik grimaced his chagrin.

'This was… disappointing.'

'I can only imagine how much so.'

Literally. And yet the spoken words could only be construed as supportive.

The other man shrugged.

'We plan to try again. There are many more cells of extremists in this country. Helpfully identified for us by MI5. And they are not all under surveillance. In the short term, however, we have no need of their assistance.'

Chasovshchik paused for a moment. Looking for a reaction. Salagin giving him nothing more than an impassive return stare.

'Questions have been asked in Moscow. There is a concern as to whether the first *Solntsepyok* operation might have been compromised.'

Salagin nodded. 'Understandably. Where victory has a thousand fathers, defeat is an orphan.'

The Watchmaker raised an eyebrow. 'Defeat? Hardly. I think we might call the result we achieved with the first warhead a dead heat. To use the language of your favourite pastime.'

And there it was again. A subtle pull at his chain. We allow you this lifestyle. The houses, the toys, the motor-racing team. And we can take this away, if we deem you uncooperative. Time, he decided, to protest, just a little.

'I presume you told them that any compromise cannot have come from within my organisation? Given, of course, the unswerving loyalty of myself and my people to the *Rodina*.'

Chasovshchik waited a moment more. Long enough that Salagin made a point of looking away from him. Casting his gaze across the ranks of desks and their occupants. Apparently untroubled.

'I have reassured my superiors that you remain as supportive as ever. And that I know with certainty that your support for the next phase of the operation will be as strong as before.'

Salagin nodded.

'You have only to ask. My entire organisation stands ready to assist.'

'As expected.'

Chasovshchik took the box off the table. A fingerprint reader at one end, waiting for Yuri's thumb. Without which the contents would be destroyed, were it forced open. Along with the transgressor's hands.

'This is for Kuzmich. It contains the essential equipment. Plus details of the target that has been selected. And a list of what is required. Vehicles, mainly. Plus resupply to the hide. We will provide you with the licence plate numbers shortly. Have them ready to deliver, and you will be informed of where, and when, in due course. The usual protocols will apply, of course.'

The usual protocols. A euphemism for no trust whatsoever. The supplies and targeting information would be driven to the hide by Yuri. The only person trusted with both the location and the instructions for the next strike. As to the cars, directions would be provided for delivery. A supermarket car park most likely. To be collected from there by the Watchmaker's men, he presumed. No further assistance needed. Salagin wondered whether the *Glávnoye Upravléniye* even needed his assistance. But knew that this was not the time to question Moscow's requirements of him.

'Of course. And we will naturally deliver to the letter of the request.'

The Watchmaker leaned across the table. Patting his hand, almost comfortingly.

'We know you will.'

10

'Golf One, Golf Three, all four targets are on the street. Probably preparing to leave.'

'Received, Golf Three.'

Mickey's finger twitched for the transmit button of a handset. Automatic reaction to hearing radio traffic after so long. James smirked at him, presumably because the urge to reply had been written clearly across his face. Mickey grimaced back at him. Turned back to the widescreen mounted on the meeting room wall.

James had reminded Mickey that the brand-new and still shiny Official Secrets Act applied to him. No less relevant, given his signature on the last version. And had clicked on the first of three icons on his laptop screen.

Mickey stared intently, soaking up the detail: 4k quality video, a clock running in one corner of the screen. Surveillance kit, state of the art. Which could only mean the war on terror. No expense spared since 9/11. Every Home Secretary since 1974 knowing that a full-on terrorist spectacular would kill their leadership ambitions. Spattering them with the sort of blood that wouldn't ever wash out.

'This is something of a collector's piece. Actual real-time

surveillance of an unfolding terrorist operation.' James stretching as he spoke. Clearly beyond weary and into properly tired. 'The problem with counterterrorism is, was and probably always will be, manpower. People cost money, and for the time being only people can do surveillance, when it comes down to it. Oh yes, drones, CCTV, facial recognition and all those other props are all very useful. But for what you're about to see, it has to be a human behind the camera. A human who, in this case, having crept into an empty house, had been defecating in a bag for the last five days. Keeping the opposition cell under surveillance in case they lived up to the chatter. The chatter indicating that they were getting ready to martyr themselves.'

Mickey had reported to the anonymous office building half a mile north of Silicone Roundabout at the appointed hour. Been processed through security. Found James waiting for him on the other side of the air lock. No handshake, for the obvious reasons. Mickey pretty sure he'd have refused in any case. Still smarting at finding himself being manipulated by a man he'd genuinely liked. Probably still did, deep down.

He searched the screen for clues as to the video's provenance. Early morning, late autumn, judging from time and light level. Therefore likely to be recent. A standard northern-town residential street. If you bought into the cliché. Terraced houses, vehicles parked on either side.

And in the centre of the shot, four men. Hanging around at the back of a car. Waiting for something. The vehicle a ten-year-old Vauxhall estate, tailgate raised. Once some salesman's pride and joy. Now just another end-of-life smoker.

Jubba jackets and kufi hats worn over jeans and hoodies. All four older than twenty but younger than thirty. Asian origin, British born. None of them wearing a mask.

'The surveillance was instigated because of their chatter. And because the cell included one of the more radical preachers in the country. Although the intelligence said there were three of them, the fanatic and a pair of fairly hapless followers. Instead of which it turns out there were four. And that wasn't the only surprise.'

The camera zoomed in. Capturing the fourth man's face for long enough that a decent hi-res image could be captured.

'Golf One, Golf Three, sending portrait of new target.'

'Golf Three, received. Designate new target as Tango Four.'

Cavendish froze the video.

'The portrait goes straight into the facial recognition database. And takes about a tenth of a second to generate an ID, *if* the target is already on record. If there's nothing on file the system then accesses the joint European database and repeats the trick.'

'And?'

'It came back with no UK record. But the Europeans had him on file. Walking out of the sea onto a Greek island two years ago.'

'Ah.'

'Quite so. *Ah.* Which means we can add a soldier to their number, of sorts.'

The camera panned. Mickey imagining the flexible lens periscoping slowly around. The street quiet. Empty, in fact. Too early for day workers, shifts already started. Nothing moving, except for one slow-moving car. Crawling along between the terraced houses, hemmed in tight by parked cars and vans. A BMW estate, clearly new.

'Golf One, Golf Three, third party vehicle approaching, index is... Charlie India Lima Eight Nine Zero Three.'

'Received, Golf Three.'

The new arrival's progress slowed, one of the waiting men beckoning it on.

'Golf One, Golf Three, second vehicle appears to be expected by Tango Four. Second vehicle has come to a stop.'

Camera zooming in. Tightening the frame to the two cars and not much else. Four men getting out of the BMW. White Caucasian, to judge from the fleeting glances that were all the camera got. Face masks and baseball caps rendering them anonymous. All moving around to the vehicle's rear. The man designated as Tango Four moving to join them. Another voice on the tape, sounding frustrated.

'I'd give my left ball for audio right now.'

A grunt of agreement.

'Golf One, Golf Three, four new suspects have debussed. All male IC1. They have opened the boot of their car. Now being joined by Tango Four.'

James clicked his mouse. The picture freezing for a moment in close-up on the terrorist.

'Tango Four seems to have been the real power in the cell. And the brains.' Cavendish tapped a file on the desk in front of him. 'Two dumb grunts and a religious fanatic to inspire them with tales of perfumed gardens and six dozen houris apiece. And that bastard to do the serious killing. Kept carefully out of sight until the last minute.'

They watched as the four newcomers lifted a heavy object

out of their vehicle's load compartment. Two men lifting it halfway out, then the other two moving in to take up the strain. Something long, and possibly cylindrical. Wrapped in a canvas sheet, obscuring line and detail.

'What's that?'

A grimace from Cavendish. 'Wait and see.'

The four men carried their cargo, visibly working hard to bear its weight. Eased it into the Vauxhall's load area with visible care. The car settling on its springs with the additional weight. Three of them returning to the BMW without speaking, the fourth leaning into the Vauxhall. Taking something from an inside pocket and placing it next to the shape. Impossible to see, bodywork blocking the view.

He worked quickly, barely thirty seconds. Then straightened up and turned to Tango Four. Handing him something the size of a small phone. Talking to him and holding eye contact. Important instructions as to how not to blow himself up. Then turned away and got back into the BMW. The four men driving away without a backward glance.

Cavendish tapped a button on his laptop, stopping the video.

'We lost the delivery vehicle, of course. The index was false, probably removed within minutes. We tracked a half a dozen similar cars and traced them all. No leads. Which means they probably made their exit without touching the main routes to avoid the ANPR cameras. Any guesses?'

Mickey was intrigued, despite his irritation.

'They moved a heavy cylinder. Which he then did something to. Getting it ready for those jihadis to use. Possibly a gas bottle, but it looked a bit too large. Something volatile?'

'You could say that.'

Cavendish tapped at the play button. Having cued up the second video. The scene flicked to a new view, shot from inside a moving vehicle. Road noise, no talk. A palpable air of tension inside the vehicle.

'Recognise the road?'

Mickey looked hard at the screen.

'I've driven that way a few dozen times for Prot. Southbound, back into London from Sandringham. M11, right? Coming up on Duxford.'

'There you go. In which case you know what's coming next. Or at least the publicly sanctioned version.'

'Shit. This is *that*...?'

'Golf Five, Golf One, status?'

The radio again, answered from within the car.

'Golf One, Golf Five, target vehicle continuing south. Speed is constant at seventy mph; driving style is controlled.'

'Received, Golf Five.'

Past the Duxford turn-off. The last chance to leave the motorway for twenty miles. Unless, of course, you knew where the undeclared exits were hidden in plain sight. Minutes passed, until Mickey was starting to wonder what the point was. Decided let Cavendish play the scene out his own way. To show no impatience, or any other emotion.

'Golf Three, Four and Five, Golf One. Decision from Operations Room. Target vehicle is to be stopped. Protocol Three has been enacted. Protocol variation follows: target vehicle appears to be carrying explosives. Target vehicle is not to be approached any closer than twenty-five metres until all suspects have been neutralised and the nature of its cargo determined. Army Explosive Ordnance Disposal personnel are inbound by air. ETA thirty minutes.'

The radio was busy with acknowledgements. Mickey raised an eyebrow at Cavendish.

'Protocol Three?'

'A long-standing tactic for stopping a target car or lorry on a motorway where there might be risk to the public. Police cars establish a rolling road block behind the target. Preferably somewhere where they can do it out of sight. Slowing the traffic to a stop. Do that and block the other side of the motorway ten miles south and you have an empty road, more or less.'

Mickey nodded. Knowing that the M11 south of Duxford would be perfect. Hills and bends restricting the view behind to less than a mile most of the time. And only two lanes, reducing the resources needed.

'And they were doing a steady seventy, so nine out of ten cars behind them and in front of the roadblock would be past and gone within seven or eight minutes.'

'Exactly. And the lorries would be a minute or so behind. Now watch.'

The cars rolled on. The camera zooming in on the Vauxhall two hundred metres in front. Four occupants staring forwards. Seemingly oblivious to the steadily less busy carriageways to front and rear. And the complete absence of traffic on the other side of the central divider. In the camera car, someone issuing orders.

'You all know how Protocol Three works. Police stop the target. Then they move in. Ready to shoot, ready to arrest. We pull up a hundred metres back to block the lorries coming up behind. Then move in to join up with them once the suspects are neutralised. We all good with that?'

A chorus of affirmatives in reply.

A white spot at the roadside. Blue lights flickering. A police car, motorway standard BMW, with an unmarked silver minibus in front of it. The police car apparently protecting a broken-down vehicle. The perfect camouflage for the stop team. Mickey half expecting any sensible would-be terrorist to pull out into the outside lane. Simple common sense, if they were inbound to a target.

But the Vauxhall stayed resolutely glued to the inside lane. Passed the BMW and was suddenly riding on its rims. All four tyres shredded by the police stinger rolled across the road at the last minute. The surveillance car breaking to a halt a hundred metres back, turning sideways. The camera panning to look through the side window. The Vauxhall ground to a stop, sitting on its battered wheel rims. No sign of any of the suspects exiting it.

Cavendish stopped the video.

'What follows was shot by the crew of Golf Four, using their drone cam. They were the rearmost of the three Security Service vehicles in the follow, and pulled up a hundred metres back from Golf Three. Tasked with overhead surveillance and stopping the lorries coming up from behind.'

He clicked on the last of the three icons. A view down the empty motorway, shot from thirty feet up. The camera zoomed out wide. The Vauxhall in the outside lane, having swerved to the right as its tyres disintegrated. The police car and van behind it and to its left. Armed policemen advancing towards the target vehicle. Helmets, tactical smocks and flak jackets. Short-barrelled carbines ready to use. The two other service cars had pulled up side-on between camera and target. Their crews out and in the shelter of the bodywork.

Movement at the Vauxhall. All four doors opening at the same moment. Obeying a command, Mickey guessed.

The camera zooming in as the drone crabbed sideways. Its operator wanting to get a better view of the scene. Zooming in tight on the suspects getting out of the car. Their hands raised. Carefully lowering themselves onto their knees, then lying flat on the tarmac surface. And as the operator zoomed out again, movement from the bottom of the frame. Police still advancing, carbines raised. The drone sank closer to the ground, moving forward behind the advancing officers.

Mickey, knowing what was about to happen, shook his head. 'Idiots.'

Cavendish shook his head. 'There were idiots involved, but it wasn't the officers on the ground. Somehow the Protocol Three variation to stay back didn't get to the people who most needed to know. Exactly why is still under investigation. They saw men surrendering, unarmed, and they did what they're paid to do. They moved in to apprehend.'

And in the last moment before it happened, Mickey realised the purpose of the apparent surrender. Watched Tango Four raise his head to look at the oncoming policemen. Waiting until they were less than a dozen metres from the Vauxhall. Saw the ingrained reflexes take over in at least one of the police officers. Rifle coming up to the aim lightning fast. Saw the prone jihadi smile, blissful in his moment of triumph. The picture vanished in a sudden searing blink of light. The screen went black an instant later.

'They found the drone two hundred metres away, in a field on the far side of the embankment. Which gives you some idea of the force of the explosion. There are ten dead. Officially nine. The terrorists, of course, and the four AFOs, practically vaporised by the blast. Plus the driver of their carrier, shredded by the windscreen as it came through the cabin. And an exceptionally unlucky Security Service team leader who had left cover to order the arrest team to get back. On top of that, one of the crew of the police car was caught in the open when the warhead went off. Bounced off the BMW hard enough to leave a dent in the bodywork. He's still in intensive care.'

'*Warhead?*' Mickey incredulous. 'You mean that was military explosive?'

'Worse than that. It seems that the men who made the handover delivered a Russian thermobaric rocket warhead to our homegrown jihadis. Originally designed to be delivered by a nasty multiple launch rocket system called Sunburn. Evidence of the Russian sense of humour, it seems. Fitted with carrying handles and with a fuse connected to a remote detonator. From the evidence recovered from Tango One's house it looks as if Borough Market was the primary target. Some sort of "do not defy the will of Allah" lightning-can-strike-twice attack. And it's been calculated that detonating that thing at the right time could have killed two hundred and fifty people, and seriously wounded as many again. Tight streets, packed restaurants and crowded market stalls. Into which those men planned to introduce and detonate a one-hundred-kilogram fuel-air device. With the capacity to heat a twenty-five-metre-radius blast zone to a thousand degrees Celsius in less than a second. And enough blast to bring down any building within a hundred metres. Proven very effective against urban defensive positions in Chechnya and the Ukraine, it seems.'

Mickey was aghast.

'But that's...'

'Outrageous? Obscene? Yes, it is. But of course the Russians would deny any involvement. They would probably claim it was a false-flag operation by an unknown third party. And then challenge us to prove that the casing fragments we've found are from their equipment. And *then*, if we could do that, to prove that the warhead came from their stocks. And not one of the countries to whom they've sold that modern equivalent of a vehicle-mounted flame thrower.'

'Which are?'

'Algeria, Azerbaijan, Armenia, Iraq, Kazakhstan, Saudi and Syria. That we know about.'

Mickey shrugged. 'They've got a point then. Or at least a decent list of suspects to hide behind.'

'Indeed they do. But they also have a Spetsnaz team in the country. Of that we're certain. Illegals, sealed off from any linkage with the "legal" intelligence apparatus in their embassy. Completely deniable, and probably not on any database. A team we don't think they've been able to extract yet. Or perhaps, and this is the frightening possibility, a team that hasn't finished its assigned task yet.'

'And you think Salagin has something to do with all of this?'

'Pavel Salagin is the last man to want to have anything to do with the massacre of a market full of Britons and tourists. Because he knows that as a naturalised British citizen, he'd go straight to prison were such a thing ever proven. For the rest of his life. But he is the sort of person who can be very efficiently leaned on to help out when the chips are down. And the chips, Michael, are most definitely down. Because we don't believe this is about a single warhead.'

He shot a glance at his watch.

'And now we have an appointment to keep. There's something we need to pick up that's going to be essential to your cover story. It's a good thing you wore the suit.'

11

The *Prizraki* were sitting around a kitchen table. Food eaten. A glass of vodka in front of each of them. A reward they had voted to allow themselves, after the success of their last operation. Relative success, of course. Not as effective as if the jihadis had detonated the warhead at the primary target, of course. Instead of which they had heard about the motorway explosion on the radio news. Instantly changing the nature of the game.

The security services had got lucky. A surveillance team had been watching the extremist group that Moscow had chosen to use to deliver the first device. Part of an operation they knew little about other than its code name. Operation Revoke.

Revoke being, it seemed, a counter-intelligence team aimed at preventing exactly such exploitation of extremist cells. Whose operations database was hidden behind a further layer of protection. And constantly monitored for intrusion. Too hard a nut even for Moscow's hardcore hackers. Who had eventually only managed to penetrate MI5's jihadist database by means of an act of treason. Blackmail succeeding where hacking failed, commissioned by an illegal agent operating out of GazNeft.

Which meant the Sunburn warhead had never been going to reach London. And made the risk of trying the same delivery method again too great. Another approach would have to be taken.

Which was why Ivan had the target list open on the screen of a tablet in front of him. Containing fifty targets across the country. All minutely analysed by the Main Directorate's strategic intelligence cell. Each location, entity or person with a detailed rationale of why and when it might be selected to be attacked. An incendiary document for the use of incendiary devices.

Ingress, egress, significant risks to the team and the recommended approach to each. Templated plans drawn up in Moscow over months of preparation for the operation. A plethora of photos, floor plans and schema for each target. Enough data for detailed planning without having to access the internet and risk detection. Projections of the damage a Sunburn could do at each location. Physical damage. Political damage. Morale damage. All compiled at the mission's inception six months earlier. And based on actual target reconnaissance, where possible.

The tablet was fitted with its own fail-safe. A thin sheet of explosive glued inside the device's plastic skin. And a fingerprint reader to be satisfied before the device could be safely switched on. The target list was open at number seven. Targeting details delivered earlier in the day, along with their supplies. Its choice unexpected, not that that mattered. A target was, after all, just that. Something to be serviced.

Ivan's team were considering their new orders. The briefing pack being discussed by the four men. Working the fine details of the mission provided by the GU's planning cell in Moscow.

Ivan's one voice among the four. Even if he was a captain among non-commissioned officers.

Sasha, seven years a *Glávnoye Upravléniye* Spetsnaz sniper. One of the shadow warriors who provided the Main Directorate with a long reach and a pitiless fist. Effortlessly gifted with firearms. Blooded and proven operating behind enemy lines in Syria. Killing ISIS leadership from a kilometre away. Unseen, reaching out and taking the lives of the enemy.

Anton, plucked from the army's Special Operations Forces. A consummate operator and experienced combat medic. Veteran of half a dozen major operations in the liberation of Idlib. Part of the unit that had rescued twenty-nine Russian military policemen from beheading. Killing two hundred of the jihadis who had surrounded them.

And Filip. Their explosives expert. As solid as granite under pressure. Another graduate of the Syrian school. Where he had specialised in the art of the booby trap. Likely to have killed over a hundred men if only a third of his carefully placed devices were triggered.

Hard men. Selected because they were the best in their fields. That, and for two other traits. Their willingness to do absolutely anything they were ordered to carry out, without qualm or question. Dedication to duty verging on fanaticism. And for the fact they were all of them completely alone in the world. Without family or close friends to restrain them. Or wonder where it was they had vanished to.

Why Ivan was their leader was still a little beyond him. The other three happily deferring not just to his rank, but his authority. Not that Ivan's authority worried him. He'd spent enough time in war zones to know that he was a born combat leader. What bothered Ivan wasn't giving the orders.

They discussed every decision in any case. Something he encouraged. What perplexed Ivan was what he was doing as an officer at all.

His path to being a *soldat prizraki* – a ghost soldier – had been very much the same as theirs. Ivan's origins being Palkino, a town in the Pskov Oblast bordering Latvia. Followed by an unexpected and abrupt military education in the Ukraine. And then Syria. Which had brought him to the height, or perhaps the depth, of his involuntarily entered profession. A perspective dependent on whose side one was looking from.

He'd already been battle-hardened by the time he joined up. Street-tough from fighting opposition fans as an FC Zenit ultra. In the course of which he'd managed to hit an opponent a little too hard in the throat. A ruptured larynx killing the man inside a minute. Advised to enlist as a means of evading any possible investigation, he'd applied to the 76th Guards Air Assault Division. Paratroopers, known to be the hardest bastards in the army. And the 76th had snapped Ivan up without hesitation. The recruiters seeing exactly the sort of pre-brutalised recruit beloved of all armies. Hoovering him up into one of Russia's most elite military units. *Vozdushno-Desantnye Voyska*. Air Landing Forces. *Desant* for short. Specialising in bringing the mobility of air assault infantry to points of strategic importance.

Parachuting, helicopter assault or just flying into captured airfields. Likely to be cut off behind enemy lines. Trapped, until the heavy armour could fight its way through to them. So *desant* encouraged independent thinkers. Junior leaders who could take command if cut off from their units. Men capable of causing mayhem behind the lines. And pointing out to the enemy that the old cliché 'I'm not trapped in here with you...' was still valid.

When it came to *desant*, the only rule was that there were no rules. Already gang savvy, Ivan took to *desant* culture like a duck to water. Comfortably able to resist the inevitable attempts at bullying. Too handy with fist and boot to be rolled over. He thrived. Won early promotion to *Mládshiy Serzhánt*. Junior sergeant. Looking set for a good career. And then, in 2014, he went to war.

An undeclared war, of course. The new normal. The entire division's fighting strength sent over the border into Ukraine. With orders to carve out and hold the region claimed by Russia-leaning separatists. To fight against men defending their homeland. Having read all about the Great Patriotic War, Ivan had pretty much thought he knew what to expect.

But even with that warning from history, he was staggered by the defenders' ferocity. Savaging the invaders at every opportunity. The 76th, deployed against the city of Luhansk, took eighty casualties in a single day. Two of them Ivan's men. One a close friend. Shot though the head by a sniper. Shrapnel from an artillery round carving a gash into Ivan's cheek. Leaving him with a wound that had healed into a long white scar. A brutal reminder of a brutal day.

When the 76th retaliated the day after, Ivan took a very personal revenge and killed for the first time. Leading his men out of their hastily dug trenches to storm the ruins of housing blasted into rubble by their tanks. Feeling nothing but pleasure in the act.

Eventually allowed leave, he had gone to visit his comrades' graves in Pskov with the rest of his squad. Disgruntled to discover that track-suited thugs were denying families of the 76th's dead access to their graves. Orders from Moscow, it seemed. To prevent journalists from reporting on deaths in an undeclared war.

Worse, the boot boys in question were a little too zealous in the interpretation of their brief. Having already given one soldier's brother a beating. For the crime of insisting he be allowed to visit the grave. Which, they had soon discovered, was a big mistake.

Ivan and his squad mates dealt out brutal retribution. Ivan himself, incensed, stamping the knee of the man he put down into screaming ruin. Then putting his knife to the man's throat. Telling him just how close he was to joining the boys in the cemetery. That done, they had paid their respects and left. Giving no fucks at all that the thugs had them on camera.

Two days later, Ivan had been pulled from the firing range by his captain. Ordered to report to battalion HQ immediately. Where he went with his face set hard. Ready for whatever shit was waiting for him there. Instead of which he'd been ushered into an empty office.

The man waiting for him was clearly former military, but wearing a quality Western suit. Introducing himself as a representative of a company Ivan was already well aware of. Knowing he was about to be sold to, rather than admonished, Ivan had put his best not interested face on. Knowing that eagerness undermines negotiation.

Upshot: the cemetery brawl had brought him to the attention of a recruiter. Who had been looking for men without either conscience or fear. With the end of his contract

Ivan mustered out as a full sergeant. And then joined the Wagner Group as a private military contractor.

Wagner, despite the Western-sounding name, very much part of the informal Russian order of battle. Mercenaries, commanded by Russian officers. All experienced men from recent wars. And recently tasked to ply their wares in Syria. To make the *Rodina's* intervention in Syria more meaningful than the occasional air strike. Going where the Russian army never could. Deniably. And without restraints.

With the rank of *Stárshiy Serzhánt*, staff sergeant, Ivan had commanded a platoon-sized force of Russian mercenaries. Quickly making them competent to *desant* standards. Using a combination of encouragement, example and old-fashioned intimidation, where necessary. Leading them in the second assault on the ISIS stronghold of Palmyra to good effect.

Taking the ground and killing scores of ISIS fighters. Suffering minimal casualties in return. Which got him promoted again, to *Praporshchik*, Warrant Officer. With a new contract, to join the intervention in Libya. As part of a force of several hundred former soldiers, again both deniable and robustly denied. Extending Moscow's influence. Earning Russia a voice at the conference setting the new lines on the map.

Ivan came home, well paid and well respected. Unsure when he'd be needed again. But quite sure it would be sooner rather than later. Back in Pskov for less than a fortnight and bored out of his mind when the invitation arrived. A plane ticket and three days' payment. With a politely worded invitation to a meeting in Moscow. Ostensibly with Wagner middle management. Details of his next assignment. Or so the invitation said.

But the meeting hadn't been with Wagner personnel. In a Wagner office, sure. But the man sitting across the desk, cigarette lit and the open pack facing Ivan, had been in uniform. Proper military uniform, not the rough approximation worn by Wagner contractors. A *polkóvnik*, according to his shoulder boards. Colonel. Looking to be somewhere in his late thirties. Young for the highest rank in the army at which anyone got blood on their hands. The most famous unit identifier in the Russian army on his arm. Blue globe inside a yellow circle, black bat symbol. *Spetsnaz*. Elite special forces.

Confused, but impressed, Ivan had taken the proffered cigarette. Kept his mouth firmly shut and listened to what the colonel had to say. The proposition a simple one. Simple, but mind-boggling.

'You are a single man, Ivan Ivanovich. You have no living relatives. You are alone in the world. And in resigning from the 76th *Desant* you left the only family that you still had. You might as well have never existed. You are not even a footnote in any history.'

The officer had leaned back in his chair, smiling. Showing every sign of enjoying the insult. Ivan had simply shrugged.

'There are schools of philosophy that believe we are all alone from the moment we're born to the moment we die. And as for existence… that's even more complicated. Are you sure you want to get into that argument?' Returning the older man's smile. 'And with respect, *Polkóvnik*, a man with your responsibilities doesn't waste his time swapping philosophies. So perhaps we can forego the mind game. What can I do for you, sir?'

The colonel had nodded. Making a decision. And not

even wasting the time needed to acknowledge Ivan's acute observation.

'I have a job for you, *Praporshchik*. The chance to do something that will poke two fingers into the eyes of an old adversary. Followed up by a good kick in the balls. And in doing so perform a great service to the *Rodina*.' He'd raised a hand, palm forward. 'No, I can't tell you what it is until you've agreed to do it. But I can tell you what it will be like.'

Ivan had nodded for him to continue.

'You, and the other three men you will command, will be alone in a foreign land. Your only support will be other agents of the state whose ability to assist you will be minimal. Limited to logistics and transport, mostly. They will be unable to get you out of any trouble that you manage to find. There will be no fire support. No extraction, at least not until the job is done. And if you manage to pull it off, you will undoubtedly be hunted. Hated. They will wish to capture you, but they will shoot you without hesitation if that looks unlikely. And you will, of course, be deniable. *Prizraki*.'

Ghosts.

He'd lit another cigarette. Had taken a deep drag and plumed smoke at the ceiling. Giving Ivan a moment to consider the word and its implications.

'We have learned a great deal in the last few years, Ivan Ivanovich. The men we sent to kill the traitor Skripal were identified by a Western research group who call themselves Bellingcat. The name is an intellectual joke, I believe. They found our men in the public record. Identified them both. Which, as you can imagine, was no more part of the plan than missing the target and poisoning a British bin-diver and his woman instead. The president was just as furious as you can imagine. And has issued strict instructions for there to be no repeat of this embarrassment. And for us to stop playing and deliver proper terror to our enemies.'

Ivan had stubbed out the first cigarette, after lighting a second from its butt.

'And so you are recruiting men without internet profiles to identify them.'

'Correct. You have never used social media, of course. Added to which your relatively inconspicuous service has already been weeded from all records. And if you accept this mission, you will enter the target state by clandestine means. Unrecorded by the apparatus of state control. Officially, you will never have existed. And we will no longer dance around the subject of revenge with poisons whose results are so uncertain. There will be no more underpant sabotage, of the type that failed so badly with the traitor Navalny. Instead we plan to use military weaponry. Warheads powerful enough to level an apartment block.'

Ivan had smirked at the grimness of what was proposed.

'You are blessed with the gift of rhetoric, *Polkóvnik*. I am already preparing myself to die gloriously in the service of the mighty Russian state, and to burnish the reputation of a...'

He'd fallen silent at the sight of the other man's broad grin.

'Perfect. That's exactly why I selected you, from all of the people that Wagner made available for consideration. When they told me that you read a lot, I *knew* you'd be a cynical bastard. And cynical bastards are exactly what I'm looking for. I don't want martyrs. And I most certainly don't want heroes. Because there'll be nothing heroic about any of this. It'll be dirty. Murderous. I need men who can carry that burden of guilt without even noticing it. And who will want to escape the enemy's traps, not go down shooting to purge their souls. But, given the reasonable chance that you will be killed in the course of this duty, I also want men who cannot be identified.'

'I see.' Ivan had taken another long drag at the cigarette. Shrugged. 'With respect, I already have a decent job. The *Rodina* already sends me to faraway places, to meet interesting new people and kill them. Just without my being formally recognised as part of the army. Why do I need to do this for you, and risk death so brazenly?'

The colonel had shrugged.

'The English talk about the stick and the carrot. It is a horse racing term. The stick is the whip applied to a reluctant mount. The carrot is the reward given for eagerness. I could simply wield the stick. Tell you that Wagner will never employ you again after today. But I am not in the habit of conscripting men. Nor do I want to have the inconvenience of all the paperwork that would result from my having to kill you. Because you would try to make me pay for depriving you of your livelihood.' As far as Ivan had been able tell from his perfectly straight face, this had not been a joke. 'So I will offer you the carrot instead, to quote our adversaries. If you volunteer for this duty, you will immediately receive the rank of *stárshiy leytenánt*.'

Senior lieutenant? Ivan had been unable to prevent his face from showing some reaction.

'Yes, I know. Having officer rank foisted on you wasn't in your plan. But it's unavoidable. Your comrades in arms are strong-minded men. They will need strong leadership, both powerful and intelligent. Leadership I expect you will be able to provide. And I will want to be sure that the men set to train you will consider you valuable enough not to damage you irreparably. And so a *leytenánt* you will be. And if you pass training successfully I'll promote you to *kapitán*. You'll be a middle-ranking officer in the finest special forces in the world. Not bad, for a man who joined the army to escape a murder charge, eh? Come back alive and successful, I'll promote you to *mayór*. And I believe the pension rights of a major are a good deal better than anything Wagner will ever pay you.'

He'd stubbed out the cigarette.

'But of course there needs to be a little stick as well. While investigating your career I found a record of a suspected crime in your personnel file from the 76th. A murder, committed just before your recruitment. Buried, of course. Like the man you murdered. Not that I blame your commanders. Who wants to waste a good man, eh? It was the one piece of evidence of your service that remains extant. Easily deleted, of course...'

Ivan had thought for a moment. On the one hand, promotion beyond his wildest expectations. And the kudos of serving as a Spetsnaz officer. On the other, presumably, a belated murder prosecution.

'But why come to Wagner to find a leader? You must have hundreds of capable men in the various branches of army special forces?'

'Yes, we assuredly do.' The colonel had shrugged. 'All of them probably known to Western intelligence services, and probably to these "Bellingcat" bastards too. Because we have until recently been too eager to rejoice in their prowess. Yearbooks. Unit photographs. And of course there are many photo-journalistic articles in the magazines celebrating our recent victories. All with photographs. All freely available on the internet, until we realised our error. Too late. All downloaded. All fed into the West's databases. Were I to pick any one of the men you suggest, I would be worried that his face might be spotted by automated CCTV monitoring. Whereas you, Ivan Ivanovich, are an unknown. I consider myself lucky that a man with your abilities has flown under the radar until now.'

Ivan had nodded. Trying not to allow his pleasure at the blunt answer to show on his face. And asked the only other question he could think of.

'And your estimate of our chances of getting away with whatever it is you have in mind for us?'

'Honestly? The best planning estimate is sixty per cent.'

'And *your* estimate?'

The colonel had shrugged again. Clearly knowing not to sugar the pill he was offering.

'Me. I'd put it lower. Sixty to forty against. But then I am a famous pessimist. Like many of my kind, who've watched the best-laid plans collapse into rat-shit on contact with the enemy. But who knows, you may drift through their surveillance like the *Prizraki* we'll make of you.'

A year later, Ivan knew that the colonel had never soft-soaped him. And that the ruthlessly honest approach had been a factor in his accepting the challenge. Agreeing to be put through accelerated *Spetsnaz* training with his new comrades. Including intensive language tuition for their target country. All single men without families. All with recent combat experience. And all enjoying the chance to gain a rank and status they never could have dreamed of.

And all, Ivan suspected, not lacking in self-knowledge, qualifying as psychopaths in their own different ways. Trading their souls for worldly possessions without a trace of guilt. Regarding it as simply the price of the mass murder.

12

'I'm told by the longer-serving members of the Service that the last twenty years have changed just about everything. At least in the world of what might be called "human espionage". Which, for the purposes of this discussion, we can extend to include acts of sabotage.'

Mickey realised that Cavendish was about to recommence the lesson. A narrative on modern espionage he'd paused when they'd left for the Palace. The box containing his medal sitting on the taxi's bench seat between them. The taxi, of course, a Service vehicle. Driven by a smirking Kev.

The box had been handed to him less than fifteen minutes previously. Awarded by the boss himself. Who had seemed just about as bemused as Mickey. His schedule thrown out by the sudden turn of events. But had nevertheless greeted the former protection officer with warmth. And a pleasing degree of restrained informality. He'd shaken hands. Enquired after Mickey's recovery from his wounds. Listening to the answer with solicitous care. Expressing pleasure at his one-time protection officer's complete recovery. Offering his congratulations on a job well done. And his commiserations at the loss of so many fellow officers. Then taken the box from his waiting equerry. Who, Mickey had noticed, had exchanged nods with Cavendish when they had entered the room. All boys together, eh?

'In the usual run of things this would wait for the next investiture, now that you're back in the country.' Mickey knowing to look appropriately abashed. The comment a small dig at his previous failure to attend for decoration at the Establishment's summons. 'But I'm reliably informed that this award needs to be made as a matter of some urgency. Not mine to question why, of course!' He'd tapped his nose at Mickey, who'd known to look both grateful and impassive at the same time. 'And so it gives me great pleasure to invest you with the medal that my grandfather instituted in the darkest days of the Second World War. It is awarded for acts of the greatest heroism or of the most conspicuous courage in circumstances of extreme danger. And I think we can safely say that you have indeed demonstrated conspicuous courage, Sergeant... ah... *Mr* Bale.'

He'd handed over the box with a small nod of respect. Shaken Mickey's hand again. Then took the small backward step that told Mickey that was that. Time to do one. And let the boss get back to whatever it was he'd rather be doing. Mickey knowing only too well what was then expected of him. Small bow, smart exit.

He shot the MI5 officer a quick glance now. Finding him gazing out of the vehicle's window as they pulled out onto the Mall. Apparently lost in thought. Tapped the box, to get his attention.

'Just before you start the next information download, why

couldn't this have been popped in the post? Or dropped on the table back in that meeting room? Why did we have to go to the Palace? And why disrupt his busy schedule?'

Which got him a raised eyebrow.

'Because, Michael, at some point your new principal is likely to ask you about the investiture. And you need to be able to answer truthfully. Given we have no idea what sort of sophisticated analysis equipment he might routinely use on his conversations.'

'Analysis?'

'Yes. Think about it for a minute. Think about how useful a man in Salagin's position will find it to know if he's being lied to. And voice stress analysis is a well-established technology. What if he asks you what was discussed? Or the circumstances?'

'And you want me to tell him the truth?'

'Absolutely. You're a former Protection Command officer who was very nearly killed. The hero who fought off assassins who were trying to kill a government minister. And you're known to the Family, from your previous job. Why *wouldn't* you get a private investiture?'

Mickey thought about that for a moment.

'Fair enough. So go on, what's changed over the last twenty years then? At a more detailed level than "everything"?'

Having a pretty good idea, truth be told, but willing to encourage the next lecture. Which would hopefully result in Cavendish to getting to the point.

'What's changed? Computing power, mostly. That and surveillance technology. It used to be the case that Service officers were set up with false identities, inserted into their target's environment and allowed to run. Lost to the opposition's view, it was hoped, in the background noise of everyday life. While the counter-espionage teams on the other side did everything possible to trip them up. But, in the main, catching them depended on mistakes by the other side, or tip-offs from informers and traitors. A bit like detectives, I suppose. But that's not really the case anymore. Have you heard of "big data"?'

Tempted to do the old 'we speak of little else in Monken Park' gag, Mickey restrained himself.

'I've heard the term. Doesn't mean that much to me though.'

'Well it's revolutionised the Security Service. And made it whole lot easier to detect and track foreign agents as they go about their business. What you do is take all the surveillance recordings, mobile phone records, financial transactions and biometrics and put them together in a great big database. Or a series of linked databases. A data lake, or a server forest, or whatever term they give it to explain it to us thickheads. Combine that with the right software to pull it all together, you can make it nigh on impossible to do anything covertly. We have facial recognition files of just about every Russian agent ever recruited, for example. And no, I don't know how. So if any of them tried to come through immigration, we'd spot them. The same goes for trying to rent a vehicle, or even make a food purchase. Because there's CCTV everywhere, and we access it all. Ditto phone conversations and locational data, And credit and debit cards. All of it constantly trawled by pattern-recognition agent bots that like nothing better than to bring their master a juicy set of connected data. All of which makes doing anything covertly nigh on impossible. Unless you can avoid the cameras. And using a phone. Or paying for anything with a card.'

He sighed.

'And yet here we are. With an untraceable team of at least four men, who have to be agents of a foreign power. Almost certainly Russia, given the weaponry involved. Although it's a reasonable guess that they're turning up the heat on the UK at the behest of their pseudo-communist partners in crime. And in punishment for suspected interference in *their* "democratic process". So, they're on the ground, here and now. Happily giving out weapons capable of killing hundreds of innocents to religion-deluded psychopaths. Driving cars that must have been provided to them by some third party. Along with the food they're eating. And they have to be sleeping somewhere, and storing the warheads too. All of which ought to be generating some sort of clues. And yet we don't have the first bloody idea as to where to start looking for them. Added to which we don't even have facial data to search for them with. Because the only video we have of them is masked.'

'You keep saying there's more than one of these warheads. How do we know that?'

Cavendish shrugged.

'We don't. But look at it objectively. They weigh over a hundred kilos apiece. So, given the means of getting them into the country would have to be been pretty heavy-lift in nature, why stop at one? And there's no way the GRU could ever be described as subtle.'

'GRU? Remind me...'

James gave Mickey a look that implied he ought to know the acronym.

'It stands for *Glávnoye Razvedyvatel'noye Upravléniye*. Or Main Intelligence Directorate. In plain speak, Russian military intelligence. Strictly speaking they've dropped the R now and become the GU. But the old version has so much traction we've never bothered noticing the change. They're the boys who tried to assassinate Skripal in Salisbury. And ended up killing an innocent who thought their nerve gas dispenser was a perfume bottle. And we strongly suspect that the four men who delivered that instrument of carnage to the jihadis of Bolton were GRU *Spetsnaz*. Their special forces. And more specifically from Unit 29155. Which is the black-ops team that specialises in assassination and subversion. And the pulling of neat little tricks like blowing up Czech arms warehouses known to be supplying the Ukraine.'

'And you're confident about all this?'

The other man shrugged. 'We're as sure as we can be. Of course it's probably not called 29155 anymore. Not since Bellingcat exposed them as the result of their generally shitty tradecraft. But then that always has been their Achilles heel.'

'Bellingcat?'

'Don't you watch the news?'

Mickey shrugged. 'I've spent the last year in the Far East specifically to get away from it.'

'And you weren't all that bothered before that, were you?' Cavendish rolled his eyes. 'Bellingcat. A private investigation group using open-source material to embarrass the Russians. Mice attaching a bell to the cat, apparently. They matched CCTV images of the Skripal kill team to the website and yearbooks of a GRU training academy. And thereby discovered that one of them was a Hero of Russia awardee. And since they were using the same first name, patronymic and date of birth as their own in their fake passports, they were pretty much bang to rights. As I believe your friends in the Met still like to say.'

'And you really think this explosive thing is more of their work?'

Cavendish nodded his head. 'We strongly suspect that the GRU has cleaned house. Set up a new clandestine subversion unit. Recruited new personnel. Cleanskins, probably, from among the little green men who invaded the Ukraine, or those private contractors who fought in Syria. Men without any records to be traced through passive means. And now they're here. And no, we don't think that one detonation was the limit of their ambitions. Because as I said, with a weapon of that weight and size, the means of transport that works for one will work for as many as you like, within limits. A shipping container, most likely. If it were my op, and I were the kind of murdering bastard that wanted to hamstring a government by starting a full-on civil war against their Muslim population, I'd ship half a dozen of the bloody things. After all, a few dead policemen was never what they came here for. So why restrict themselves to one shot?'

'And can we prove any of this?'

The MI5 officer shook his head. 'No. And we don't need to. All we need is to find them. At which point special forces will be turned loose on them. With all the fire support they might need to do the job, up to and including helicopter gunships. The latter being on permanent warm standby until they're found and killed. Ready to lift off in under fifteen minutes and fully bombed up. So you can believe me when I tell you that when we find them the gloves will come off. And stay off until they're all either captured or dead. Preferably with some proof of what they were up to.'

'And you think that Salagin is involved?'

Cavendish's turn to shrug.

'As I said earlier, we very much doubt any direct involvement. But *someone's* providing them with assistance. Hiring them cars. Buying them food. Perhaps even providing somewhere to hide, and store the warheads. And with your new employer probably being one of the best-resourced people in the country, with a strong loyalty to the Russian leadership, he's of great interest to us.'

'Guilty until proven innocent?'

'Very much so.'

The MI5 officer turned a hard smile on Mickey. And *there* was the real James Cavendish, Mickey realised. The man who'd gone after assault-rifle-armed killers with only a pistol.

'Just find them for me, Michael. I'll do the rest.'

13

It was always going to be Albie who asked. Sammy knew that all too well. He'd known it ever since he'd come clean to his fellow gang leaders the previous year. Admitted to using an assassin to deal with their colleague Joe Castagna. And had been expecting Albie to come calling sooner or later. Counting on it, even. Every month he hadn't more of a surprise than the last. Having bet himself that Albie wouldn't last six months without asking to play with Sammy's Action Man.

Which meant he'd been amazed when a year passed without any such request. And almost relieved when he got the call from Albie a few days later. A call out of the blue. Offering hospitality. A neutral venue, of course. Plenty of notice, allowing Sammy's boys to make their investigations. Concluding that if Albie were planning a hit on him then it would be MAD. Mutually Assured Destruction. Suicidal. And so public as to be very bad for business indeed. And so Sammy had gone along with the meet. Already knowing what Albie wanted.

Drink was taken. Bread was broken. Pleasantries were dispensed with. Albie's gift of vintage scotch opened. Tasted and declared excellent. Knowing that Albie could dance around the point in question for another half bottle, Sammy got to the point.

'What can I do for you, Albie? Not that it's not good to have a drink with a colleague...' For which read *competitor*, in truth. Their territories overlapping, which resulted in a never-ending push and shove. A healthy friction, as both men knew. Darwinian in its ability to weed out the weak. And to bring the strong to their attention, for that matter. To be nurtured and moulded. Or quietly removed, if they looked too strong for comfort. 'But I can't help thinking you din't come here just for the pleasure of my company.'

A point Albie conceded with a curt nod. Hard to deny, really. Given he was known to look down on anyone who wasn't a white male of a certain age. Sammy, of second-generation Chinese immigrant stock, a second-class citizen by Albie's standards. To be tolerated, and nothing more.

'You got me, Sammy. Thing is, I've got this problem.'

Sammy raised his eyebrows. Encouraging Albie to keep talking.

'Not like you to have any such thing, Albie. You being the least tolerant and most feared of all of us.'

Albie had inclined his head in thanks for the respect. Which had been genuine, as far as it went. Albie always having been the least willing of their tight group of gang leaders to tolerate a slight. Just not minded to actually go looking for disrespect he could punish. That being the one small but glaring fault that had led to Joe's permanent retirement.

'There's these blokes... kids really, putting the arm on me. Ex-army. Got their five years in, learned which end of the stick goes bang, now they think they're all that.'

'And they have the weapons to back that threat up?'

Sammy, suddenly very interested. Knowing that whoever threatened Albie today would be at his throat tomorrow.

'Looks like it.' Albie took a sip of his drink. 'Lovely drop of gold watch that. Yeah, I lost two blokes last week. Disappeared. For a while.'

'And now they've reappeared?'

'Right. And they wasn't in the best of nick, neither of them. Given some fucker had given them a right brassing up with a machine gun.'

'A *machine gun*?'

Albie sniffed. Like he was making light of it. Ignoring the slight tone of disbelief in Sammy's voice.

'Tell me something else that puts a dozen bullet holes in a geezer and I'll call it that instead.'

Sammy nodded slowly.

'That is bad news.'

'Yeah. Innit? Because they'll be moving in on you next. Unless we do something to stop them.'

'And you're thinking I can help?'

Sammy knowing exactly what Albie was thinking.

'Yeah, well… that geezer you mentioned in the caff. Last time we met, yeah? The one you told us you'd used to…'

He raised a meaningful eyebrow. Clearly expecting Sammy to fill in the blanks.

'Bale,' Sammy interjected. Not wanting their former associate Joe's name spoken out loud. A hope doomed to disappointment.

'Yeah, that was it. Ex-copper, right? The one that topped poor old Joe Castagna.'

The poor old Joe in question having been the biggest nutter in North London. A gang leader without scruple, hesitation or indeed restraint in dealing out havoc. Whether necessary or not. Who became so dangerous for all concerned that Sammy had taken unilateral action. Used a contact inside Joe's gang to feed information to Bale. Whose exceptional abilities with firearms and his fists had done the rest. He'd mounted a campaign of vengeance for his dead sister. Conveniently killed by Joe's drugs. A campaign that had resulted in Joe's emphatic demise.

And for Sammy that had been enough. Job jobbed. The threat that Joe's behaviour had posed neutralised. His territory parcelled out between the rest of them. And the whole thing to be quietly forgotten.

Except by Albie. Of all their fellow gang leaders the most impressed by the gun-for-hire thing. Sammy had seen it in his eyes. When the truth of his chosen executioner's rampage had been revealed.

'Bale. You're sure? He won't be easy to wake up. And he'll be a bastard to control if we do. Last time I had his need for revenge to manipulate him with. This time he'll be a wild card. You sure you want to deal that card out?'

Albie raised his eyebrows.

'You got any better ideas? There's two of them, hardcore. Twins. Half a dozen of their mates hanging on for the shits and giggles, but they'd fold quick enough if the ringleaders were gone. I reckon your man could disappear them without them ever even seeing him.'

Sammy smiled slowly.

'He *really* ain't gonna like it.'

'I... *we*... don't need him to fucking like it, Sammy son. We just need him to do it.' The older man leaned closer, taking a pair of silver tubes from his pocket and unscrewing their caps.

Sliding out a full-sized Cuban cigar into Sammy's waiting hand. The blissful scent of fine tobacco curling around his olfactory sense. 'I can provide him with the weapons and the money. And the intelligence as to where the fuckers are living. And, most importantly, the *motivation*. I just need you to tell me where I can find him.'

Sammy pondered that.

'Better if he doesn't know how you came by his name. Leaves my leverage on him unspoiled. Just do what you normally do to motivate the reluctant, eh? It ain't like you don't enjoy it?'

Albie smiled.

'Problem with you, Sammy son, is you know me too well.'

14

Mickey reported for his orientation at the town house the next day. Dressed as advised by Kuzmich. Yuri Kuzmich. Salagin's head of security. A man whose every aspect epitomised darkness. Saturnine in both appearance and speech. His voice deep and with a distinctive rasp. Where Salagin's was light, a relative conversational rapier, Kuzmich's was a lead-filled cosh by comparison. Black clothing from head to foot. Top brands, but strangely devalued by the absence of colour. As Mickey decided to make a point of not remarking on his first morning. Welcomed to what Kuzmich informed him was known as UGS. Shorthand for Upper Grosvenor Street.

Mickey, of course, wearing muted colours. High-quality suiting. Good black shoes. A restrained tie. All of which carried the tag 'obviously' for Mickey. Smiling when Kuzmich had bluntly detailed his expectations. Proving Salagin's comment about the man being spot on.

He'd purchased three identical top-end suits for the gig. Dunhill, his brand of choice now that he had money. Courtesy of Joe Castagna's fallen empire. One to wear, one for spare and one for the dry cleaner. Dark grey, a wool and cotton blend for comfort. With a midnight blue check so subtle as to be no more than a hint. Tastefully stylish. Subtle enough to fade into the background. And bloody expensive, by Mickey's usual standards. Spending his money like a man who knew it could be confiscated at any moment.

He'd added half a dozen white shirts to the bill. Top-end fabric, 140 grams per square metre Egyptian long-staple cotton. Mainly just because he could. Plus half a dozen smart but sober ties, Tom Ford for the most part. Because that was how Mickey rolled. When the money was unrestrained, that was.

Footwear was tempting, but already sorted. Three pairs of Loakes from the Prot days. Already broken in. Comfortable, practical, repairable. Which left only one thing missing. Pointed out by Kuzmich during his close inspection of Mickey's get-up. His own suit tailored to be slightly baggy around the left chest. Not dramatically obvious, but enough to accommodate a slim handgun. Something small calibre or with a single-stack magazine, Mickey mused. Old habits dying hard even in the post-Prot world of no guns at all.

'You miss the Glock, I think. Protection Command carry the Model 19, yes?'

Mickey agreed. Both with the model number and that he missed it just a little. Ignoring the slightly ham-fisted attempt to prove credibility. No harm in the head of security thinking he was cleverer than the new guy. And no point in denying the slight feeling of nakedness either. Working protection without a weapon like going to a party without a bottle. And so he nodded a rueful acceptance of the point. The word from Cavendish being that Kuzmich was a man to be careful around.

'You probably don't remember Boris Yeltsin, do you, Michael?' Cavendish had asked earlier.

Mickey had felt a little insulted, truth be told. Like James thought he was thick. Rather than just not cut from the same cloth.

'Russian president after Gorby, right? The bloke who did such a good job of liberalising Russia that they went straight back across the road to dictatorship when he'd finished.'

Cavendish had raised an eyebrow. Whether at the attempted humour or the accuracy of the comment not clear.

'In a way. He did open Russia up, but an awful lot of monkey business took place during his term in office. Anyway, Salagin's man Kuzmich was a member of the Presidential Security Service that guarded Yeltsin. Selected for his exceptional service in Afghanistan as a junior KGB officer. With something of a reputation, forged in the most difficult part of that god-awful war, towards the end. So don't whatever you do make the mistake of underestimating either Kuzmich's efficiency or ruthlessness. His boss Korzhakov was Yeltsin's protection officer, back when he was the party boss in Moscow. And when Yeltsin was dismissed, Korzhakov resigned from the KGB and the Communist Party. Set up a unit of former KGB officers, including Kuzmich, to guard him from assassination attempts. He was in place, with Kuzmich at his side every step of the way. Which made Salagin the beneficiary of enough inside information to make himself very rich indeed. In the years when Yeltsin spent most of his time emptying vodka bottles and giving away bits of the country's industries to anyone that asked nicely enough while he was under the influence.'

'How did Salagin come into it? Did they serve together?'

Cavendish had tapped the file in front of him.

'Nobody seems to know. There's certainly no record of any military service by Salagin. He seems to have been enjoying a life of small-time crime when Kuzmich gave him the call. Our best guess is that the two of them grew up together in the same benighted Moscow suburb that made Kuzmich so tough. And between them, with Kuzmich's links into Yeltsin and Salagin's business instincts, they pretty much cleaned up.'

'But they eventually found themselves having to leave the party?'

'Indeed. Korzhakov got himself sacked. For overstepping the limits of his power. At which point Kuzmich and Salagin made a hasty exit. Probably astute enough to see that the whole thing could only end in tears. That, and the likelihood of a bigger bastard coming along and taking their newfound wealth off them. Clever of them, but then as a former KGB officer Kuzmich would have known enough about Putin to see him coming, I suppose. And by the time Salagin decided to quit Moscow for London he was easily worth ten billion. Small fry when compared to the US internet moguls, of course, but powerful enough. That power being largely exercised on his behalf by Kuzmich. In their Russian heyday that meant breaking up strikes, and making sure competitors didn't get too big for their boots. Anything that would benefit from the application of a little muscle, that was where you'd find Kuzmich. He might have a comparatively easy life now, but he doesn't seem to have softened one little bit. Keep a watchful eye on him, Michael, because he's where the bulk of the danger lies.'

'Salagin's just the front man then?'

Cavendish had nodded. 'We strongly suspect so. Certainly Kuzmich behaves much more like the head of an intelligence outstation than a commercial head of security. Although that might just be old habits dying hard.'

Waiting for him at the staff entrance, Kuzmich gestured to the airport-style X-ray machine and metal detector that dominated the space. Reached by a flight of stairs that led down from street level to the first basement level. A pair of uniformed security officers stood behind him, smirking at Mickey. One man, one woman. Making sense with so many female staff to process every day. And making Mickey suspect that the first-day initiation James had warned him to expect was par for the course.

'So, Michael Bale. Show me what you have brought with you into Pavel Ivanovich's house.'

Mickey decided to play it dumb. 'Pavel who now?'

The Russian raised an eyebrow. 'You British. Every time I think I have reached the depth of your indifference to other cultures... Pavel *Ivanovich* Salagin. His name is Pavel. His father was Ivan. Therefore Pavel, son of Ivan. Iv-an-o-vich?'

Rolling his eyes. Mickey shrugged artlessly. Secretly delighted to have given the impression of being sandwich short so early.

'Gotcha. And as to what I have in my pockets...'

He put his wallet, keys and phone on the table. Kuzmich looked at them and shrugged. Gestured to the security staff for a lock box.

'None of this is allowed into the house. They will be stored until the end of your duty period.'

Mickey raised a hand. 'The wallet and keys are fair game. But the phone is important.'

Kuzmich shrugged. 'The phone is the most dangerous of the three. It will be locked safely—'

'No. It won't.' Mickey, folding his arms with just the right degree of obduracy. 'For one thing, I'm a trained royal protection officer. And the Queen didn't expect me to put my phone in a box when I booked in for a shift. Ask me why.'

Kuzmich stared at him wordlessly. His expression unreadable.

Mickey shaking his head in disgust. 'Because, in case you hadn't guessed, I'm not a security guard.' Flashing the two suddenly glowering uniforms an unbothered stare. Thinking in passing that the female half of the partnership would probably scrub up pretty well, out of uniform. 'I'm a professional protection officer. I don't take calls on duty, but I do retain the phone because I might need to handle a domestic emergency.'

'Domestic emergency? You live alone. Your wife left—'

'Yes. *Thank you* for that reminder. I do have parents though, if you look hard enough through your database. And they're both old enough for a domestic emergency to be a distinct possibility. You can examine it all you like. But I'm keeping the phone.'

Kuzmich looked at him for a moment. 'There are many other protection officers available on the London market, Michael Bale.'

Mickey shrugged. Moment of truth. Go big or go home. Quite fancying the going home option, truth be told. Although his natural urge to win was flicking two fingers at that sentiment even as his rational side entertained it.

'Yeah. So you go and find one with my experience. With a bit of luck you might even find one with a long service medal.'

But no fucking George Cross, buddy. So stitch that. Reaching out to retrieve his wallet, keys and phone. Hoping that Kuzmich understood the reasons why Salagin was so keen for him to join the team.

'Wait.' The Russian levelled a foreboding stare at him. Then turned and hooked a thumb over his shoulder at the two uniforms. 'Out.'

He waited until they had left the room. 'You will not speak to me in such a way again if you wish to keep this job.'

Mickey shook his head. 'You get back what you give in this life, Mr Kuzmich. Treat me like a wanker and I will return the favour. You understand *wanker*, I presume, since you're not indifferent to other cultures?' He picked the phone up. 'And now I'll be on my way. You can tell Mr Salagin that—'

Kuzmich raised a hand. Not looking happy. Swallowing his pride, perhaps?

'We will compromise. You will provide me with the phone. I will have my security expert check it. After that you may keep it. But you will absolutely not use it on duty unless it is a *domestic emergency*. Agreed? And it will be checked each time you come on duty, and when you leave. To make sure no inappropriate calls have been made.'

'Agreed.'

Doing his best not to look smug. Suspecting that he'd failed. Kuzmich held out his hand. Waiting until Mickey took the unsubtle hint and placed the phone on his palm.

'Dimitri!'

A late-twenty-something poked his head around the corner of an office door. Specs, tatts, long hair, neckbeard. Lacking only a Cyberdyne Systems T-shirt to be the archetypal hacker nerd. Probably clever as all hell. The guy the techs at the Box outstation would be jousting with, in skills terms. Although

Mickey would be the one watching the point of a lance come spearing towards him.

'Dimitri will check your phone on arrival, and on leaving. *Every* time. Scan it to ensure it has not been tampered with.'

Mickey looked into the office. The usual hacker paraphernalia. Three screens. One running a real-time from the X-ray machine. Mickey seeing his coat pass under the all-seeing eye. Kuzmich gave him the phone. And, Mickey was amused to see, he visibly sneered at it. *Not flashy enough for you, youngster?* There was a weakness to be exploited.

'Check this very thoroughly. Mr Bale will surrender it to you for examination every time he reports for duty. And when he leaves. You know what to do.' He turned to Mickey. 'Very well. You will need to understand the layout of the house. I will walk you through it, while we wait to find out whether you have "malware". Pay attention, there *will* be a test.'

He led Mickey out into the house.

'This is level minus one. The highest of the basement levels. This half of the floor is for the servants.'

One side of a wall that divided the floor into two was utilitarian in decor. Clean, smart even, but no more. A quartet of cubicle-sized bedrooms for anyone whose duties made it necessary or sensible for them to stay over. Each with a capsule bathroom. Then the kitchen. Occupying a squash court's footprint. And presided over by a formidably talented chef. A close-cropped bottle blonde. Intricate ink vanishing up the rolled-up right sleeve of her whites. Her every word clearly that of a beneficent despot. Her staff drilled to perfection. Working on lunch for twenty with brisk efficiency.

'We can feed one hundred, with additional agency staff.'

Mickey nodded, looking around and taking mental notes.

Impressed by the pristine cleanliness and order. And by the aromas competing for attention. Noting the cargo-sized lift in one corner.

Kuzmich led him through a door. Key-carding them into another world. The pool's water so clear as to appear invisible. Only a slight ripple as it was drawn into the filters. And the pool itself Olympic length. Illuminated by hidden uplighters. Their glow seemingly as natural as daylight. With woodwork that could have graced a Riva speedboat.

On the pool's far side, one last door. The cinema. Or rather, Kuzmich clarified, the larger of the two. Thirty lounger seats arrayed in two shallow arcs. Facing a genuine cinema screen. Carpet and wall decor designed to kill echoes. Present the best possible soundscape. Another lift at the rear of the room. Which Kuzmich ushered him into. Pressing the down button.

'This is Pavel Ivanovich's toy cupboard.'

They stepped out into level minus two. Mickey as close to a double take as he'd ever been. Saved only by his experiences in the royal household, over the years. Not that the Boss had anything this. There being two halves to the garage. Cars for show, and cars for go. The former a who's who of automotive aristocracy. Two Ferrari 250 GTOs, one blue and yellow, one blood red. Both in exquisite condition. Porsche 917, original Ford GT40, Aston Martin DB5, 6 and 7. Plus half a dozen classic F1 cars. One for every decade since the sixties. Individually every one of them a famous car. Iconic. Collectively literally priceless.

The cars for go an equal revelation. Absolute top-end state-of-the-art automotive royalty. Lamborghini Sián Roadster. Porsche Carrera GT. Ferrari F12 Tdf. Pagani Huayra. Bugatti Divo. Then the high-end limos, a Merc S-class in pole position for the vehicle lift to the street above. A Range Rover offroader for the country. An Audi S8 for grunt and a Rolls-Royce Ghost for waft. All new. All gleaming.

'Impressive.'

Kuzmich sniffed. Seemingly not sharing Mickey's reverence.

'Pavel Ivanovich invests with the eye of a collector. Come.'

He led Mickey back to the rear of house lift and pressed Zero. Ground floor. The two men emerging into a gleaming gymnasium. Mirrors aplenty, air con cool. Twenty or so fitness machines. And a personal trainer sitting reading in the far corner.

'Pavel Ivanovich likes to exercise. I insist he does so in here, of course.'

Mickey nodded. Very much approving. They went through the male changing room. Into the house proper. Defined as such by the quality of the flooring and fabrics. Kuzmich pointed to a door on the right.

'Dining room.'

Mickey opened it, casting a swift glance around the tables.

'This is the public dining room?'

'Of course. He spends most of his time upstairs.'

Mickey's approval as to the security arrangements increasing. He closed the door and followed Kuzmich into the main reception room. Dignified and understated, but opulent all the same. Staying true to its stated purpose. Enough room for a reception for as many dinner guests as could be fitted in next door.

'I see it's excess all areas.'

'Excess? You pronounce access like a Russian. This is deliberate…?' Kuzmich stared at him. The penny eventually dropping. 'This is a joke, I see. Very funny.'

He led Mickey through into the front of house. Reception, and the sitting room in which he'd met Salagin. More security, suited and booted. Mickey noting the presence of half a dozen wardrobe cupboards beside the doors. Guessing that one or two of them were for taking things out of, rather than putting them in.

'The firearms are all licensed?'

Kuzmich followed his stare. Nodded. Looking ever so slightly impressed. 'Of course. And if the cabinets are opened an immediate alarm is sent to the local law enforcement. The weapons are for use only in the last resort.'

Mickey kept a straight face. Pretty sure that any alarm from this house just wouldn't be summoning the local nick's relief. Kuzmich gestured him into the front-of-house lift. Which, doors closed, took them up one floor. Depositing them into a featureless corridor ten metres in length.

'Not quite what I was expecting.'

'This way.'

He led Mickey halfway down the corridor. Turned to the blank wall and waited. After a moment a hidden door hissed open. Revealing another lift.

'Activated by retinal scan. We will add your data to the security system later. Perhaps. The rest of this floor is reached from the one above it only by stairs. And the walls of this corridor are sheet steel. Backed with Kevlar and then more steel. Anyone looking to use this route into the inner household would not get far.'

'Impregnable.'

'Yes. There is another entrance, via the backstairs. A motorised steel door. With an emergency close function. An intruder would need a cutting torch and a lot of time to open it.'

Kuzmich got into the lift. Beckoned Mickey to join him.

'There are only a very few who know of this. And the non-disclosure agreement you signed is very specific—'

'—as to the penalties, legal and financial, for revealing it or any other aspect of the household to any third party. I noticed. When I read it before signing.'

The Russian shrugged. 'The NDA would be the least of your problems.'

Mickey shrugged off the threat. Not intending to break the terms of the NDA. Unless left without any choice in the matter. Were his unwritten contract with MI5 brought to bear. MI5 pretty much trumping any threat Kuzmich might have to offer.

'Cleaners?'

'They are blindfolded on their way into the house. It is contractually specified.'

'I'm sure that makes it all feel perfectly normal.'

Kuzmich shrugged.

'They earn a very good rate in return for their discretion.'

'Female guests?'

The Russian looked at him levelly. 'There is a line in an American gangster film that reflects Pavel Ivanovich's approach to life. To always be both willing and able to walk away from your life in an instant. Without any warning, if necessary. And he has found, over the years, that female guests tend to bring a certain expectation with them. Which has led him to restrict his romantic activities to a simpler basis.'

Meaning, Mickey presumed, that the female guests were also paid well for their discretion. Which was not life as Mickey would have chosen to live it. But each to their own.

Kuzmich pressed a button, and the lift ascended a floor.

Disgorging them into a lobby whose degree of opulence was another step up from the public floors.

'Three floors. Eight bedrooms. Each one with living room and bathroom. A larger living room and dining room, plus the usual.'

Mickey, slightly amused. 'The usual?'

'Electronic games room. Old arcade games, mostly, plus racing simulators. Whatever is desired. Snooker room. Sauna and jacuzzi. A smaller cinema. Library. The usual.'

'You've been away from normality too long, Mr Kuzmich.'

The Russian nodded. 'Possibly so. But I too could walk away from all this at the drop of a hat.'

'You live here?'

'I am Pavel Ivanovich's head of security. Of course I live here. I have a flat on this floor. And now for the test.'

Mickey grinned. 'Go for your life.'

'How many people can be fed in the main dining—'

'Thirty. That was the number of places set. Although those tables had the look of extendables. So I'm going to guess forty-eight.'

Kuzmich nodded. Making a point of being unimpressed.

'How many catering staff were there?'

'Eleven.' The Russian drew breath to reply, but Mickey raised a finger. 'Or at least eleven that you let me see. But I could hear two more washing up as we walked through from the gym. They were talking. So let's call it thirteen, eh? Any more, or am I boring you yet?'

Kuzmich grunted, starting his tour of the family's inner sanctum. Later, phone returned, Mickey had his retinas scanned. The device apparently having been proven innocent by whatever analysis it had undergone.

'This means I passed the test, right?'

The Russian nodded dourly. 'You may return home. I will email you with your work pattern once I have discussed it with Pavel Ivanovich. Be ready to attend at 0800 tomorrow morning.'

15

Mickey was almost home when Albie picked him up off the street. Two minutes from the apartment block. A big car pulling up at the kerb ahead of him. A dove-grey Rolls-Royce Ghost. The Extended version. With enough room in the back for a game of squash. He clocked the car's index number. GAN 805S. Gang boss? Cute.

The nearside passenger door opened. A suited heavy climbing out. Clocking the street for any sign of police. Then opening his jacket to show Mickey the butt of an automatic pistol. Mickey stopped walking just inside touching distance. Confident that he could close to knife-fighting range before the thug could get the pistol out of the holster. Or just land a very hard kick in the balls.

'Very nice. Did you get the bullets that come with it for Christmas too?'

The big man's face hardened. Mickey filing his sense of humour failure under weaknesses to be exploited. And waited on full alert for any reaction. Instead of which a head craned around the car's door pillar.

'Don't be a cunt, son. And get into the motor. Nobody says no to *me* twice.'

Mickey took stock of the car's occupant. Late fifties, early sixties perhaps. But heavily creased. Excessive consumption of alcohol and tobacco, at a guess. Quality schmutter, probably Savile Row. Its excellence somewhat undermined by actual snakeskin boots. And gold. Lots of gold. A thick neck chain. Several chunky knuckleduster rings, probably last used as such in the twentieth century. And a gold Rolex President so unfashionable as to be cool, *if* it had had the right dial colour. Mickey suddenly very tempted to put his boot heel through the veined red nose. Already instinctively knowing what was going on.

'No, don't tell me. I know it...' Put a hand to his forehead as if straining for the answer. 'I've got it! You're an associate of a deceased London gangster, right?'

Which got him a pained smile. 'Fucking spot on, buddy. Now get in the car and no unpleasantness needs to befall anybody. I only want a little chat. We'll drop you back here.'

Mickey looked at the other man for a moment. Then at the bodyguard. Who seemed to have been modelled on a badly stuffed sofa. And decided not to bother going to war yet.

'If I get in the car, what's the odds I end up encased in concrete?'

The Roller's passenger grinned. 'Nah, nothing like that! I just want to talk. And I want to do it away from flapping ears.'

Mickey shrugged. Getting into the car. Knowing that his location was being tracked by James's operational analysts. And easily enough tied to local CCTV. The bodyguard climbed in beside him. Drew the automatic and rested it on his knee. Making Mickey shake his head, if only metaphorically. Pigeon-holing the man into the 'gun-dependent' category.

He mentally rehearsed the attack. Turn and put a knife hand into GAN 805S's throat to keep him busy. Using the move to give him stand-off distance on the mug. Then land a precision-guided backhand knuckle-fist to gun lover's temple. Following up with the waiting right. Bouncing the mug's head off the armoured glass behind him. Applying further concussive violence if necessary to subdue him. Then taking the gun and putting it up under the older man's sagging chin. Since he suspected firearms were a language the gang leader would speak fluently.

Then waited, ready to strike. The car pulled silently away from the kerb. Probably as hushed as the reading room at James's club.

'So you want to know who I am. Don't you, son?' Mickey nodded reluctantly. Not really all that interested, truth be told. But hoping to encourage some getting to the point. 'You, Mickey, can call me Albie. I was a colleague of Joe Castagna's.' Mickey gave him a blank stare. Albie notching up the irascibility. 'So don't take the piss, and don't try to waste my time. I know who you are, Mickey Bale. And I know what you done to Joe.'

Old news then.

'And?'

A grin spread across the older man's face. 'Yeah, you fucking *have* got balls the size of melons! Sammy said you'd front up like you hadn't a clue and didn't give a shit either.'

'Sammy?'

'Sammy Chin. Geezer that put me onto you. And more to the point, the geezer that persuaded you to top poor old Joe.' Albie nodded knowingly. ''Course you think it was a bloke calling himself Nemesis, but it weren't. He was just the puppet. Whereas it was Sammy pulling his strings.'

Mickey digested that. Decided to stay with the no-comment response. 'I don't know—'

'Let's make this quicker and easier, eh? I know what you did to Joe. I know what help you had. And I know there was money involved. A *lot* of money. See, we inherited his empire, between us. The families divvied it up. And I got the books.' He gave Mickey a significant look. Raised eyebrow, hands spread for effect. 'No comment? All right. So I had the books audited, and it turns out there was millions not accounted for. Quite a lot of millions. And there's only one person that could have had it away with all that money. And the assets involved. And that's *you*, son.'

Assets. Drugs. Diamonds. Bitcoin. All confiscated by Mickey and his partners in crime. The freelance assault team who helped him take Joe down. The price of their participation.

Mickey smiled bemusedly. Gesturing to the car's opulent interior. Serenity seating, heated, cooled and massaged. Fridge. Starlight headlining.

'So I've got this man Castagna's money? And yet you're the one cruising around in a quarter million pounds' worth of Roller. Nice colour too. While I work as a bodyguard to put bread on the table.'

Albie shook his head. Affected sadness with the attempt to gull him.

'Nah. You en't fooling me, son. I know you live in a shithole flat, but that's just deception, innit? A couple of years, you'll vanish off to the sun and properly live it up. Or that *was* your plan. Until your uncle Albie showed up to show you the error of your ways.'

'And now my plan is…?'

Albie grinned again. Shark-like. Very likely to be the last smile some people ever saw.

'Well that's *better*! No more pointless denials. And very wise of you. Because I know enough about *you*, Mickey my son, to have you sent away for a very long stretch indeed. And while you were away, you could have spent many a happy hour imagining what might be happening to the people you care about. Your dear old mum and dad. That tasty little ex-wife. Without you around to protect them.'

Threats to his nearest and dearest being something that Mickey was discovering didn't lose their impact. Or the reaction they inspired in him. The urge to homicide. Albie's voice hardened. As he hoisted himself into the metaphorical saddle.

'Your new plan, Mickey son, is to be my *fucking* bitch. To cough up the money you took from Joe. *All* the fucking money, right? And to put yourself at my disposal. For whenever I want someone dealing with in a terminal manner. After all…' Mickey guessed what was coming next. 'I know what you did to some of Joe's closest associates. If you can get to them, I reckon you can get to anyone. You come and work for me, you'll live well. I guarantee that. And I got a job for you to start with, as it happens. Some former squaddies who left the army with more than they took in, if you know what I mean?'

Mickey stared at him blankly.

'Guns, son. They're tooled up to the eyeballs with automatic weapons. And they're making, shall we say, demands. Stronger demands than I feel inclined to let them get away with. Only problem being that I don't have that sort of muscle in my corner. Which is where *you* come in.'

'And if I decline?'

Albie smiled beatifically.

'We both know you en't going to do that son. Because I reckon you still got feelings for your ex-missus.'

Mickey affected to think for a minute. Wondering what James would be making of all this, if he was listening in.

'I'll give you one chance to back off. Because I work for some people who'll go through you like you're not even there. And they won't appreciate you trying to hijack my services.'

Albie sneered. Determined not to take no for an answer. And omnipotent, in his head at least.'

'Nah. You ain't working for any of the families. I done my checks, so as not to tread on anyone's toes. Besides which there's no other bastard in London that could be any sort of threat to me.'

Mickey sighed. Put on his 'you saw right though me' face. *On your own head be it.*

'All right. Say I play along. It'll take me a few days to get the money. And there won't be as much as you're hoping for. I had to share it with the people who helped me.'

Albie grinned triumphantly. 'See, Keith? I said you wouldn't have to tickle him up, didn't I? No, son, I *know* you only kept a share. Stands to reason, dunnit? Other palms to grease. People to respect. But what's left is *mine*. All you have to do is give me access to your bank accounts. The secret ones. I see any large sums transferred out, I'm going to assume you're using the money to put a hit on me with your special forces mates. And that, son, ain't gonna end well for you or anyone else I connect with you.'

Mickey raised an eyebrow. 'You think I'd need help? I'm insulted.'

The gangster laughed, loving the bravado. While completely ignoring it. 'What, you think you're enough to take me down? Nah, son, it'd take a fucking airstrike to get me. I'm *very* well protected. So, are you going to be a good boy for your uncle Albie?'

'What do you think?'

Albie smirked. 'I think you're going to bend over and take it dry.' He pressed a discreet switch on the door next to him. 'Here'll do, Gerald. Mr Bale can walk to a tube.' Handed Mickey a card. 'I'll give you forty-eight hours. Give me a call inside that window, tell me you're ready to do what's expected, and it'll all be sweet.'

Mickey got out of the car. Forced to squeeze past Keith. Who deliberately only half opened the door. And who was clearly doing his very best to imprint himself on Mickey. Who in turn resisted the urge to sweep-kick the mug's legs out from under him. And then give him a fast lesson in what happens when a mixed martial arts fight goes to the floor. Let him get back into the car. Then held the door open and fixed Albie with a direct stare. Ignoring Keith reaching for the pistol.

'Very well, Mr Ward. But when this all goes sideways, just remember that you came looking for me. You wanted this.'

Albie gestured to Keith to close the door. And Mickey looked around. Guessing that he was in the middle of Friern Barnet. Twenty minutes' walk from the nearest station. Pointedly ignoring Keith's amused smile through the armoured glass. Watching as the Rolls-Royce wafted away. And waiting for his phone to ring.

16

Mickey found a caff over the road from Arnos Grove tube station. Got himself tea and a slice. Read a book on his phone until James arrived an hour later. Bought himself a cup of tea and sat opposite Mickey. Eyes watching the street for any hint of surveillance.

'There's no doubt as to what this man Ward wanted from you?'

Mickey pursed his lips and shook his head. 'Gang money and dead rivals. In that order. I have to call him before this time two days from now. Or bad things start happening to people I care about.'

'I see. And do you have the money he wants?'

'Nowhere near all of it. And most of what I do have is inaccessible, given Ian Shaw is holding it for me.'

The ex-special forces operator whose team had helped him put Joe Castagna away. And then grassed him up to James without any hesitation. Not that Mickey would have done anything different in Shaw's shoes.

James raised an eyebrow. 'No ace in the hole? No walking away money?'

Mickey shrugged. 'I do have a little something left that I kept from Castagna's stash. It's sitting in a safety deposit in town. But it'll sting like hell to hand it over to a gangster like that.'

'I don't think the money's the issue, Michael.' James sipped at the tea again. 'Ye gods, I swear they brew this stuff for hours rather than minutes. No, the real question is what are we going to do about his demands for you to perform executions for him?'

'Yeah.' Mickey drank, looking at James over the rim of his mug. 'Thing is, he mentioned some ex-soldiers. Said they're tooled up with weapons they seem to have borrowed from their former employer. Which sounds like the sort of thing you lot would want to stamp on?'

James laughed softly at his hopeful tone. 'I doubt MI5 could care all that much about civil policing issues like that. Even if they do involve the use of stolen military weapons. No, I think we're on our own with this, at least for the time being. My suggestion is that you take whatever money you've put aside for a rainy day to Mr Ward, and get him to tell you exactly what it is that he'd like you to do for him. We can worry about the detail later.'

'That's the second time you've said "we" inside a minute.'

James smiled into his disbelief. 'Whatever endangers you endangers our mission. Which, you might recall, is all about making sure that the Russians don't manage to smuggle a Sunburn warhead into a sports stadium or a shopping centre. So if it's all the same to you, I'm in.'

Mickey shrugged. 'It's your funeral. What does MI5 do to officers who indulge in a little bit of extra-curricular activity?'

'I have no idea. But since I've already been drummed out of the only job I ever really loved for the crime of too much intelligence, I doubt it'll really bother me too much.'

'OK.' Mickey shrugged. 'I haven't got a clue as to how to deal with Ward though.'

James raised an amused eyebrow.

'Given the exploits I've been reading about from your crusade to take Castagna down, the one thing I'm not worried about is your ability to come up with something. And then to do whatever it takes to see it through.'

17

York was quiet at one in the morning. A Monday, chosen deliberately to minimise the number of people on the streets. Not because collateral damage wouldn't be a good thing. The more people caught in the blast the better. But Moscow had chosen this target for them with a very definite objective in mind. It wasn't death, or even destruction that was the main driver. Indeed a conscious choice had been made to avoid the more obvious targets, where 'collateral damage' would be much higher. But whose impact on the mission's eventual success would be nowhere near as great. The GRU's strategists had decided to provide MI5 with something else to focus on. A classic *maskirovka*. A word meaning to mask one's true intention. And in this case, to misdirect the men hunting the ghosts.

The two cars drove in towards the target with a sixty-second spacing. Beyond covert. A pair of anonymous rentals. Estate cars, carrying a Sunburn warhead apiece, seats folded down. Vehicles procured for them by cut-outs. Who had bought them using forged ID. Albeit ID with perfect entries in the government's driver licensing database. Entered by a member of staff for whom working from home was perfect.

Enabling her activity to go unnoticed. Fake licences, but squeaky clean if checked. And with another pair of equally untraceable cars parked close by. Insurance, in case they needed to break surveillance. In the event of some sort of contact.

Ivan was in the lead vehicle's passenger seat. Scouting ahead. Making certain that their path to the target was clear. And what he saw gave him no reason to abort. Empty streets. Pubs closed. A lone policeman heading in the opposite direction from the target. His stride purposeful. Probably heading for his mid-shift meal.

Both men put on their face masks, to remove any CCTV risk. Ivan's driver Anton taking the turn into College Street. Driving slowly and silently on into Minster Yard. The car almost silent on battery power alone. A steady cruise down a quiet street. Deserted. A dead end. Harder for a swift extraction but minimising the risk of exposure to chance observation.

Ivan got out, leaving Anton at the wheel. Walked to the car park gate. Looked at the cottages on the other side of the road. Dark, without any sign of life. He took a slim aerosol from his pocket. Used it to give the gate's hinges a generous spray of lubricant. Then sized up the padlock. Finding it just as the reconnaissance team had pictured it. A simple enough commercial model that he could have picked, given a minute or so. No need. A master key had been provided in the mission equipment envelope that matched the target's number.

The gates opened noiselessly on freshly oiled hinges, and he beckoned Anton to drive in. His teammate three-pointing the BMW to face the gate. Ready to exfiltrate. Then got out and joined Ivan at the gate. Both men listened, watching the cottages across the narrow street for any sign of activity. Still nothing.

'I say we commit. You concur?'

Looking at Anton questioningly. No rule said he had to do so, of course. But Ivan knew his men appreciated being consulted. A luxury for a Russian NCO – consultation. So often just expected to clean up messes caused by poor decision-making.

'I concur.'

Ivan broke squelch on his radio twice. A slow double click that told the following delivery vehicle to proceed. Four rapid clicks being the message to disengage. Return to the hide without looking back. Both men turned back to the looming stone edifice. The plan sensibly ignored the target's imposing western and southern entrances. Likely to be under close CCTV surveillance. Their ingress instead planned to take place through a small modern annex. As they approached it Ivan realised that he could hear the faint strains of organ music. Probably loud, inside the building. Likely to be plainly audible, if they opened the door. Anton tapped his ear, then pointed at the door. Whispering a question.

'Abort?'

A good question. If the door were opened, would the noise released alert the inhabitants of the houses? Was someone's late-night practice about to frustrate their plan? The answer came almost immediately. The music stopping. A few seconds' gap before it started again. And stopped. And started again. The same few bars, repeated. The organist seeking perfection.

Ivan shook his head. Pointing at the small door set in its stone arch.

'Unlock. Do not open.'

The other man nodded, going to work with his lock-picking tools. The lock giving up the unequal struggle within seconds. The radio clicked once, and Ivan turned to see the second vehicle had arrived. Filip and Sasha. And the device.

'We proceed. I go in; you wait. Two clicks means bring the device.'

The NCO nodding and going down on one knee in the shadows. His own machine pistol ready for use.

Ivan waited for the music to stop. Opened the door and slipped through it into a storeroom. Closed the door quickly. The music started again, louder now he was inside. Not religious music, something classical. Whoever was playing it probably lost in the joy of the magnificent instrument. The door on the far side of the room was unlocked, and Ivan went through it into the cathedral's dimly lit interior.

The music was loud now. The organ's power almost palpable in the air. The unseen organist throwing everything into their performance. Ivan paused for a moment in the shadows. Partly to enjoy the music. Partly to acclimatise his eyes to the massive building's gloom. And partly to give time for anyone else in the cathedral to reveal themselves.

A forest of organ pipes vanished off up into the gloom. A single spot of light beneath them showing the organist's location. The organ's choir console, according to the manufacturer's website. Approaching silently between the choir stalls, Ivan stopped a few feet from the instrument. Taking a moment to appreciate the player's virtuosity. How long would it take to learn that level of competence. A decade? Two? A lifetime? Had the man labouring to create such a glorious noise been playing since childhood? So much expertise, the labour of decades.

Shrugging off the momentary hesitation, he coughed. Loudly enough for the player to hear him. The other man stopped playing. Turning on the console's bench in surprise. A cleric. Wearing what he believed the English called a dog's collar. Fifties, slightly overweight, half-moon spectacles. Soft hands, soft eyes. Threat assessment: none.

'You play well. What was the music?'

Baffled, the clergyman answered him without thinking.

'Bach. Fantasia in C Minor, as it happens. But you can't be in the Minster at this time of...' His admonishment tailed off. Silenced by the sight of Ivan's machine-pistol as it came up from his side. 'Really, we don't have anything worth—'

Ivan shot him. Once, though the forehead. The gun's report through its suppressor no louder than hands clapping. Painting the organ console behind him with a spray of blood. Waited for a full minute in silence for any sign of anyone else in the massive stone building. Then went to help the others with the warheads. They carried the Sunburns into the choir one at a time. Their two hundred and thirty kilo weight enough to make the four men strain in lifting them. Placing them in front of the organ console. None of them sparing more than a swift, dispassionate glance for the dead priest sprawled across its keyboard. Collateral damage.

'Set the detonator to remote.'

Filip nodded, readying himself to visit dragon's breath on the ancient church's timbers. Taking the detonator from his backpack. Mating it to the warheads with a pair of cables. And tapping in the long numeric code required to arm the device. The warheads in their optimum location, according to their briefing notes. Close to the middle of the cathedral.

The place in which their explosive power would cause the most destruction. And none of them wanted to be anywhere near when they detonated. Sasha looked up at the magnificent vaulted ceiling.

'Will the explosion collapse the building?'

The explosives expert answered without interrupting his work. 'Completely? Unlikely. It will blow the central tower apart – that's certain. But the weight of the stone in those walls will probably be too strong for even this much punch. Of course it will blow the roof off and the windows out. And incinerate anything that's left.'

Instantaneously destroying an architectural marvel that took hundreds of years to build. Ivan shrugged, shaking off his admittedly practically non-existent guilt.

'Are they ready?'

Filip stood. Nodded. 'They will detonate when I send the signal.'

'Very well.' Ivan gestured to the side door, hidden behind the choir's benches. 'Let us change the game.'

18

Niall Kerr was having problems sleeping. Which was nothing new, admittedly. But particularly bad that night. A combination of too much caffeine and not enough cash flow. Keeping him awake into the small hours. From experience he would catch some sleep from around two o'clock. But wake again before six. With nothing changed.

His coffee shop was on the verge of going under. Crushed under the weight of business rates, minimum wage and government loan repayments. His personal bank account, credit cards and store cards all still deep in the shit. Repossession the only possible next step. Probably starting with the house. Heart-breaking, given the amount of equity he had in it. Knowing that the various financial institutions to whom he was indebted would pillage it mercilessly to pay off their recovery costs. Liable to leave him with nothing to show for fifteen years of payments. Or even still in debt, and homeless too.

Something caught his attention. A click. Soft, barely audible. But still there. Someone breaking into the Minster? He got out of the bed he wasn't sleeping in and went to the window.

Two cars in the cathedral car park. One just starting to roll, the other waiting behind it. Niall watched from behind the lace curtain his ex-wife had put up in happier times. Before she decided to stop tolerating his slowly ratcheting tension. The first car eased out through the gates. Running on electric power. Considerate. The same with the second. Decent of them not to start noisy internal combustion engines.

He leaned forward. Watching as the two vehicles moved slowly down the narrow street. Both with two occupants apiece, dimly visible in the light cast by their dashboards. Which struck him as odd. Why bring two cars when one would have been enough? He stared out into the street's pitch-blackness for a long moment, then shrugged. Turned back to his rumpled bed. Perhaps another try at sleeping would bear more fruit.

The Sunburn's fuel-air fireball lit the room like a floodlight. Shining through the cathedral windows like a star fallen to earth. A blink of light bright enough to read small print by. Niall knew he would recall the irregular pattern of bumps and marks on the old house's wall behind his bed for the rest of his life. As he turned to see what had illuminated the room an immense bang slapped the house. Followed an instant later by a hail of lead and glass shrapnel from the intricate stained-glass windows. Blowing off roof tiles and crazing the window glass. Blocking his view of the conflagration inside the cathedral. Other than the ghostly image of a huge, orange-lit cloud rising from within the ancient building.

The light dimmed. The fireball inside the ancient church's walls starting to fade. And Niall's stunned mind began to come to terms with the devastation. A small part of him just starting to exult. Thanking his good sense in renewing his building's insurance. Feeling a weight lift from his shoulders. Realising that the house would most likely be a write-off. Leaving him with enough money from the equity to start again.

And then that blissful prospect suddenly, terribly, became reality. The house's destruction no longer theoretical. A roof timber, blasted into the air by the force of the explosion, came spinning back to earth. Punching down through Niall's roof. Pulverising the bedroom ceiling. Taking his life with the contemptuous ease bestowed on it by mass and velocity. Momentum momentarily granted to it by the product of a Russian weapons factory. And hammered his smashed body down through two floors before coming to rest in the cellar.

19

Mickey's phone rang. Waking him from a deep sleep. Picking it up from the bedside table, he looked at the display though half-opened eyes.

02.45. What?

Caller ID: *Jimmy C.* Mickey's little joke.

He considered ignoring the call. Realised that James wouldn't be pocket-dialling at three in the morning. Thumbed the green icon.

'It's two forty—'

'Switch on your TV. Pick any news channel – they've all got it. Call me back once you're awake.'

The call dropped, leaving him staring at the screen. What the fuck? He staggered out of bed. Considered coffee. Decided against it. Knowing it would prevent any further sleep. Sat down in front of the TV and pointed the remote. Selecting BBC News 24. Stared at the screen for a moment in disbelief. Real-time footage from a drone. Night vision camera. A building that could only be a cathedral, given the dimly discernible cruciform shape. The lights of fire engines illuminating portions of its bulk. And its roof almost completely gone. The central tower no more than a stub. The windows empty gaping holes where intricate glass had been. Glass now glittering in fans across the cobbles. Debris scattered across the open ground around the massive building.

The news ticker announcing that York Minster had effectively been destroyed by an explosive device. No confirmation so far that the device used matched whatever it was that had exploded on the M11. Sleep clearly unlikely, he made a coffee. Then called James back.

'Tell me that wasn't a Sunburn.' Silence at the other end. 'Jesus wept.'

'We think it was more than one, given the devastation caused. Could have been worse, in one respect. They could have smuggled them into a shopping centre in Manchester, or Birmingham, or Leeds. They could have killed hundreds. Instead of which they've wrecked an old church. Probably killed a few people in the vicinity, but still light casualties compared to what might have been.'

Some old church. And Mickey pretty sure that being one of the 'light casualties' still made you just as much of a casualty.

'There must be a rationale.'

Amused irritation in the reply. 'Of course there's a rationale. Wait until you see the headlines tomorrow, if you're not sure what it is.'

Mickey, watching the ticker, saw it change. A phone call taking responsibility received by several media outlets. On behalf of something called 'Fire of Islam'.

'Hang on... they called it in?'

'We're not convinced. There's no record of any such jihadist group in the UK or globally. The calls were made on a satellite phone with location finding switched off. And the message was the same every time. We're betting that once the analysts have had the chance to take a good look at the recordings they'll find it's the exact same message being replayed.'

Mickey thought.

'False flag?'

'You tell me. But I can guarantee you that there's going to be a chorus of outrage tomorrow. And some pretty strong opinions as to what the Service is doing. And what more needs to be done. With New Britain calling for internment, I'd imagine.'

'I see.' Mickey yawned. 'What do you want me to do about it?'

'You're on duty with Salagin at 0800?'

'Yes. First day.'

'Good. Take the phone with you. It's time to up our game and see what we can find out.'

20

Mickey arrived for his first shift ten minutes early. Having got an Uber in from Monken Park. Keen to avoid the risk of a tube delay putting first-day mockers on him. Handed his keys and wallet to the security guard on duty. Watched as she put them into the mini-safety deposit box. Which was then slotted into the hole it had been pulled from. Put his coat into a plastic tray. Pushed it into the X-ray machine. Walked through the metal detection arch. And took his phone from the coat pocket.

'Hey! Dimitri!'

Mickey being a firm believer in starting training early. He waited for the hacker to appear at the office door. Handed him the phone. It had already been scanned once before, of course. With nothing untoward found. Because, sensibly enough, it had been nothing more than a phone. Nothing more offensive on it than that klaxon ringtone he'd been meaning to delete for months. But which he never heard, because only Roz's number would activate it. Installed as a joke with Roz, in happier times. Probably still there because he wasn't admitting to himself that she was gone yet.

The phone had been returned to him by Kuzmich at the

end of his tour of the building. The Russian sneering at the device as he handed it back.

'Your phone, Michael Bale. It seems you have a very boring life. No online gambling. No adult dating. What *do* you do for fun?'

I kill drug dealers and spy on Russians. As Mickey was tempted to reply. Satisfying himself with something a little more anodyne.

'You know. Takeaways and box sets. The odd spot of bag work when I feel like breaking a sweat.'

Kuzmich had looked baffled. The concept of a punch bag outside his cultural boundary, it seemed.

'*Bag work?*' Bag pronounced beg. Mickey scenting faked incomprehension. 'What is this *bag work?*'

Taking a half-step back as Mickey had raised clenched fists and took guard.

'Punch bag. You know, a hanging bag for punching exercise? I used to box a little.'

The Russian had grinned. Sizing up the three-inch advantage in height and probably in reach that he had on Mickey. And the twenty kilos in weight. Mostly muscle.

'Ah! A new English word. This is good. I will let you know the next time I need a *punch bag.*'

The phone Mickey handed to Dimitri wasn't the same device, of course. Even if it had been scuffed to match the original with meticulous care by Service technicians. The screen corner minutely cracked in perfect imitation. Hiding its additional capabilities in plain sight. Mickey waited while whoever was in the office took a good look at the X-ray. A fair precaution, he supposed. Made small talk while he waited. Smiling at the security guard.

'I owe you an apology.'

She shot him a sideways glance from the screen she was looking at.

'Why's that, *sir*?'

Sir with a 'c', perhaps? Shots fired, to judge from her tone. He grinned. Employing a little of the 'little-boy Mickey' charm. The act that Roz had used to rip the piss out of him for. Back when they were married. In the years BC. Before Castagna.

'I was somewhat disparaging, when I came in for my first meeting with Mr Kuzmich.' Treading carefully. For all he knew she and Kuzmich were an item. And so obeying the first rule of not knowing a situation. Walk on eggshells until you know what's really underfoot. 'And I apologise. Unreservedly.'

Mickey always careful never to '*want* to apologise', having once been floored by the reply 'go on then, get on with it'.

After a moment she looked up at him. With a questioning expression. Mickey sensed there was more apologising to do.

'It was unworthy of me. I came across like a right prick. And even if it was because he provoked me, I had no place disrespecting you and your mate. So I'm sorry.'

The smile broadened.

'Your apology is accepted. Although the truth is that neither Gerry nor I could hold a candle to your medal.'

A slight hint of piss-take? Mickey jumped eagerly on that opportunity.

'The GC? Truth is I was unlucky. Then lucky, then unlucky again. And that's about all there was to it.'

Her head tipped to one side. Interested. Willing to accept him back into the category of decent enough bloke, perhaps.

'Go on.'

He shrugged. 'Unlucky to be on the spot. There were five

hundred of us it could have been, but I had to be the one to win that lottery. Then I got lucky in having a former SAS major in the car with me. He did most of the heavy lifting.'

'And the second unlucky? Getting shot yourself?'

He shrugged. Tapping his abdominals with a finger.

'This? No. The bad luck was losing my mate. He got shot by one of the assassins. DOA.'

Her gaze softened a little. And Mickey wondered if using Wade's memory as a tool to unpick her hostility was too cynical. Taking a moment to offer his respects to his fallen colleague. The loss showing on his face, to judge from her sympathetic tone.

'I read about it. A lot of good coppers went down that night. You were lucky to survive. Or just good.'

Mickey smiled wanly.

'Bit of both, perhaps.'

Her eyes lost focus for moment. Someone talking to her through the earpiece, perhaps. She raised a hand and went into the security office. Mickey heard voices. Mostly Dimitri's. She came out with his phone and a pissed-off expression. Clearly not entirely happy with whatever the hacker had said to her. Muttering something Mickey might not have been intended to hear.

'I swear one day I'm going to fuck that dirty bastard up so badly...' She pasted on a smile. 'Your phone came up clean. Here you go.' Handed it back to him. 'Mr Kuzmich is waiting for you in the front lobby. I'll tell the other boys you're not quite as much of a prick as you showed out the first time you were here.'

Mickey inclined his head in recognition of the concession. 'High praise. See you later... ah, I didn't catch your name.'

She studied him coolly for moment. 'That's because I didn't give it to you yet. But you can have it, because you apologised nicely. I'm Angela. And you are…?'

Knowing his name. Just not which version he'd like her to use. Michael for formal. Mickey for casual.

'Mickey. Good to meet you, Angela.'

He walked through the house, finding the big Russian sitting behind the reception desk.

'Mr Kuzmich.'

'Michael Bale. Your duties today will be to escort Pavel Ivanovich. He has a busy schedule. Here.'

He passed Mickey a sheet of paper. Several appointments. Starting with an 0900 meeting at GazNeft in Southwark. Mickey nodded. Folded the paper and put it in his pocket.

'Mr Bale.' He turned to find the man himself stepping out of the lift from the inner household. 'Already I feel safer.'

Mickey returned the smile. 'Mr Salagin. Already I feel better off.'

Kuzmich came around the desk. 'Today you will drive Pavel Ivanovich. I will accompany you. You recall the operation of the vehicle lift?'

'Of course he does. The man is a professional.' Salagin switched the smile's gentle point of focus to his associate. 'We will use the Mercedes, yes?'

'The keys are in the car. The night staff have cleaned and refuelled it.'

Mickey led the way to the lift. Opened the door and ushered his new principal and Kuzmich into it. Taking a slim metal rod and a flat metal rectangle from his pockets.

'This is essential equipment?'

He nodded at Salagin's question as they left the lift. And

ignored Kuzmich's sardonic gaze. Walking through the motor-sport collection to the everyday vehicles.

'It is where I come from. There are no car-washing facilities here. Which means your vehicle has been off the premises overnight. Meaning that I have to assume a potential risk to you. Don't worry, this won't take long.'

He affixed the mirror to its handle and opened out the telescopic rod, which was four feet long. Switched on the mirror's built-in light and put it beneath the car. Checking all four wheel arches, then the car's underside.

'It's clean.'

He opened Salagin's door. Closing it once Salagin was seated. Got into the driver's seat. Kuzmich already sitting in the front passenger seat. Disdaining his seat belt, Mickey noticed. Also noting the lack of any protest from the car. Deducing that the seat belt alarm for that particular chair must have been disabled.

'You were expecting a problem?'

'No.' Mickey, collapsed the mirror's handle and put it back into a pocket. Then turned back to face Salagin. 'But it's a risk that takes a minute to cover off.'

The Russian looked at him for a moment. 'I knew you were a serious man. A professional. But there are limits to your ability to protect me, are there not? You are no longer armed.'

Mickey nodded. 'The law takes a dim view of unlicensed firearms.'

'In which case you will need to decide how to deal with the fact that there is such a weapon in this vehicle. Open that armrest.'

Mickey looked down. Then pushed the button that lifted the leather covered halves of the storage box. The pistol was clipped into some sort of housing. Secured by what looked like a fingerprint scanner. Kuzmich grinned at his expression.

'Place your right forefinger on the glass.'

He did as he was bidden, and the mechanism clicked. The weapon released, springing up a centimetre or so. Looking like the Glock Mickey was used to, but a shrunken version. He bent to look at the slide. Reading the model number. Noting the spare short magazine alongside it.

'G26 Gen 5?'

Kuzmich reached into the box and brought the stubby pistol out. It looked like a toy in his big hand.

'It is smaller than the usual. Made for concealed carry. Which also means that it fits the storage compartment.'

He put the weapon back into the clips. Then pushed it down, to secure it. Mickey wiping the fingerprint reader with his shirt cuff.

'You're making something of a presumption, gentlemen. And taking something of a risk, legally.'

Salagin shrugged. 'As to the presumption, to some degree, yes. I am presuming that, in the event of an attempt on my life, you would prefer the chance to shoot back. And as to the legalities...'

'Don't worry my pretty head?' Mickey frowned. 'I thought your associate here seemed a little zealous. But this...'

Salagin raised an eyebrow.

'Most of my other bodyguards have been delighted. Like boys in a sweetshop.'

'Most of your other bodyguards have probably never killed. Or been shot. Neither of which I'm in any hurry to repeat. If fact I'd bet that most of them were never even properly trained to use one of these.' He gestured to the stubby weapon. 'The short barrel makes it a close-quarters gun. Twenty metres or less, if you want to do more than just spray and pray. And the stub grip will make it a bastard to keep on target.'

Kuzmich nodded approvingly. 'Michael Bale knows his firearms; this much is clear.'

Salagin nodded. 'And now perhaps you know why I am willing to pay the oleaginous Colonel Smythe what he wants to secure your services? My previous guards drooled when they saw that gun. You, on the other hand, are more... thoughtful.'

'And that's all there is? Just the one weapon?'

Salagin had the good grace to look slightly guilty. 'The boot lid might contain something with a little more punch. Don't look if you don't wish to know.' He raised his arm, tapping the watch under his suit. 'And now, I think, we need to be leaving. It would not be good for me to be late to my meeting.'

21

James took his place at the meeting table. Keeping his head down, metaphorically. Determined not to speak unless he was spoken to. Knowing that in this company he was a beginner in a very grown-up game. There because he just happened to have the phone number for the Hereford Sports and Social Club. That, and a special friend on the inside with one of their suspects.

Around the table: Susan Miles, team leader for Operation Revoke. The hunt for a presumed team of Russian GRU Spetsnaz operators. And the missile warheads they were using.

Anthony Harding, her boss. The Service's deputy director general. DDG for short. Reporting to the director general. The DG's focus tending to be on strategy and politics. Meaning every DG needed a deputy with a rat-trap fierce brain. To run the operational side of the shop. And make sure that critical ops didn't go wrong. Or at least not badly enough to inflict blowback. Harding being that fearsome intellect.

Miles's three team leaders filled the remaining chairs. Web surveillance. Intelligence. Field operations. All eyeing each other and their bosses with a wary expectation of unhappiness.

'So... Susan?'

Harding putting the ball squarely on his subordinate's plate. Quite right, of course. Her operation, her leadership. But also sending a very clear message. Given that Op Revoke really wasn't in a good place. That Service senior management were calculating when to close the blast doors. And minimise risk to both the Service and their political masters. In that order, naturally. Facilitating the identification of the perceived failures. Thereby enabling swift and decisive management action. To learn from mistakes made. To improve future performance. And to select the necessary scapegoat.

Of course Susan probably had a lot more rope left yet. At least judging from the operation room's shop-floor gossip. Neither DDG nor DG being the types to make needless changes mid-operation. Or to be pushed into premature action. But when Susan reached the end of that rope the jerk would be violent. She looked around the table, meeting each of their gazes momentarily.

'Very well, let's summarise what we do know. Then we can get into what's still unknown, as of now. Summary: a large explosive device, probably a pair of daisy-chained Russian Sunburn warheads, was used to destroy one of the country's oldest religious monuments early this morning. The second explosion in what now looks like a planned campaign whose ultimate objective is, as yet, not completely clear. Although not too hard to guess at. We believe that a single member of the clergy may have been murdered prior to or in the explosion, since the cathedral's archdeacon is missing. With another four dead and seventeen wounded and injured. And we know that a previously unknown apparently extremist group calling themselves the Fire of Islam have taken responsibility through calls to several media outlets. And not that it's our problem, but the far right is going berserk. Probably gaining members faster than a gym after lockdown. What can anyone add to that? David?'

David Slater. Heading up a team of fifteen web analysts. Teams of five, working around the clock. Sitting at the heart of the Revoke team in a tight huddle. Constantly trawling both the internet and the dark web. Both the surface and the murky depths of the internet. Looking for any way to work out what known and suspected terrorist threats were up to. Using a slew of false identities to infiltrate their networks. Drawing the line only at outright entrapment. Probably.

'Nothing conclusive. As we're already aware, the known cells have all been woken up, one at a time. Lots of chatter, most of it encoded. The occasional careless message by text, but nothing to indicate anything other than the desire to act. Not the intent. It's like they're all waiting for something. And nobody's talking to anyone else, which implies that they've been warned to keep it tight.'

Susan nodded. As expected. And deeply worrying. Implying external control of the UK's extremists. 'Intelligence?'

Katy Spinner. Everyone's pick for the DDG slot, fifteen years in the future. Sharp as a razor. And equally hard. Her team not a formal component of Op Revoke. A temporary secondment. The result of the DG's demand for prompt results. Wielding her team of bright-eyed analysts like a scalpel. Working to peel away layers of deliberate and coincidental confusion and obfuscation. To get to the bright shiny truth of the matter in hand. Her people using algorithmic agents to trawl a range of databases. Whether they had formal access to the data or not. Constantly updating their understanding of the never-ending secret war with the UK's rivals. With little distinction between friends and enemies. Slicing, dicing, collating and presenting the ever-changing analysis. Helping Susan's team understand what they were dealing with.

'The same as before.'

Susan's reply a study in neutrally stated disappointment.

'Nothing new then.'

Not even a question. Knowing that if there was anything to work on, it would be on the table. Katy shook her head.

'Nothing concrete.' Counting the points off on her fingers. 'There's still no trace of the weapons entering the UK. My guess would be either a freight container, hidden among the thousands unloaded every day. That or a charter flight. Whatever method they used to get them in was almost certainly previously unused. Other than perhaps for an unloaded dress rehearsal.'

She tapped a second finger.

'As to personnel, there's also still no trace of the delivery team coming in either. Assuming they're foreign and not domestic. Of course a really ballsy way of doing it would have been to bring them *and* their weapons in at the same time. Which almost certainly means a flight. We're examining radar recordings and cross-referencing to flight plans and passenger manifests, although a decent-sized twin jet could have bought in that much weight without any trouble. And we have at least one target with access to a small airliner.'

Salagin, of course.

A third finger.

'As to the events of earlier today, CCTV from around the Minster gives us two cars, both two up. Occupants masked, plates faked, and the approach must have been carried out off ANPR routes. Because that CCTV is the first time we see either car on the way in, and the last on the way out. There's no physical evidence of entry. No fingerprints. No footprints. And we suspect they used a DNA masking spray as they extracted from the cathedral.'

Traces of a strong alkaline compound having been found around the presumed point of entry. Not that there'd have been much DNA left in any case, by the time the fire brigade had finished putting the fires out. Katy still talking, tapping finger number four.

'Turning to general counter-intelligence, we've asked the agent runners to shake the illegal tree long and hard. But none of the usual sources are admitting to knowing anything.'

Illegals. Russian and former Warsaw Pact sleeper agents for the most part. Smuggled into the UK decades before under false identities. Undercover operatives positioned to carry out acts of espionage and sabotage. Their raison d'être suddenly ceasing to exist along with the USSR itself. Reactivated in the late nineties by the new Russian regime. Those who hadn't already returned home. Now approaching the end of their operational capability. But still potentially useful. Having been tracked down by the Service, over three decades. And brought into the light, one by one.

Easily encouraged, in some cases, by the certain knowledge that their cause had failed. And then replaced by a kleptocratic dictatorship. Others less keen, wanting only to continue the innocent lives they had assumed were their reward. And some having to be dragged, kicking and screaming. Co-operation under duress, with threats of long prison sentences. While for a few outright bribery was all it had taken. But whatever their motivation, none had been involved in hiding or supplying the bombers. Katy tapping finger five.

'We've also requested a particularly careful watch on the legals. But if our friends at the embassy or their friends at GazNeft are involved, they're doing it without giving us any clues.'

Legals. Spies whose presence in the country was officially sanctioned, albeit under the cover of non-spying roles. A long and honourable pretence on the part of all intelligence services. And therefore tolerated by all. Avoiding tit-for-tat expulsions for the most part. And giving everyone a window into what everyone else was doing. Helping to deflate mutual paranoia. Although James suspected that GazNeft's nest of spies was somewhat beyond that brief. And that the Service would like nothing better than to have the Met kick those doors in, sometime soon.

'And no joy getting anything out of GazNeft?'

Miles knowing the answer. But needing to be seen to be thorough.

'No. It's the same old story. The top two floors of their building are intrusion proof. They use window vibrators to make laser mikes useless. And no way to get in there and implant anything because they keep it buttoned up so tight. Although if there are GRU operatives among their number, their signal discipline is impressive.'

Security so thorough, James had learned, that the floor below the lower of the two GazNeft levels was rented out, but empty. Patrolled by security guards from the floor above, preventing any intrusion activity from below. All incoming and outgoing utilities subject to real-time jamming of piping and cables. And all maintenance carried out in-house. By carefully vetted contractors, Russian émigrés to a man.

Their loyalty either already total or assured by their familial links back home. In most cases both.

Susan nodded her head. Her expression grim. And moved on to the last report.

'Field Ops?'

Tom Stoddard. Late forties, grizzled, weathered. Not a high-flyer, but nobody's fool. His bike jacket and helmet bespeaking his everyday engagement at the sharp end of the operation.

'We've got the square root of bugger all. We're keeping a loose watch on the most likely jihadist suspects. All the usual, covert cameras, discreet follow jobs, but we're coming up empty-handed. All of these terror groups waking up at the same time means we're stretched to the max watching would-be bombers going to get their groceries. All innocent, it seems. The Burnley Two cell was wiped out in the M11 explosion, and no-one else is giving us any reason to step up to proper intrusion. Not that a judge would sign off on, at any rate. So right now I've got nothing. And if you want my opinion, we're being misdirected. Made to look in the wrong place. And I think this latest attack is part of it.'

Susan turned to James. Who was, all things considered, very much her Hail Mary play.

'Liaison?'

James kept it crisp.

'Our asset in the Salagin household went in hot for the first time this morning, so it's a bit early to say, really. His recruitment went smoothly enough though.'

The DDG leaned forward. 'This is the former policeman, Bale, I presume?'

Sounding not entirely approving. James making his face immobile as he answered. 'Yes, sir.'

'And you'll vouch for him?'

James stared back, not entirely sure what the senior officer was asking him. But suspecting the surreptitious easing of armour plate under backside. Sensing the interest around the table. DDG and the new boy, seconds out, round one. He decided to play it completely straight. If a little irascibly.

'I'll vouch for the fact he took a bullet in the guts to keep the current Home Secretary alive a year ago. Not forgetting the assassin he took out while he was gut shot. Sir.'

Subtext, unspoken but acid-washed into James's tone: do you think you have *any* right to doubt the man's loyalty? Harding, to be fair, took that right on the chin. Not much choice, seeing that James had stopped a round from the same assault rifle. But straight up and down of him all the same.

'Ah. The George Cross thing. Yes, point taken, Major Cavendish. I suppose what I was really asking was whether he's any *good*?'

James raised an eyebrow. 'As a spy? He's probably as completely out of his depth as I would be. And he'll either sink, swim, or just splash about so much that the opposition, if that's what they are, will have to get rid of him. If only to spare him the embarrassment. So I think all we can do is wait and see how he does. But given the criminal acts he's *alleged* to have carried out prior to the assassination attempt, I'd say he's a cool enough head.'

Harding nodded.

'Thank you.'

The meeting ran to a close. Nobody happy, nobody giving up. All knowing that what they needed was a lucky break. To find the loose thread in whatever plan was being perpetrated on UK soil. And then pull, good and hard. Susan made a brief speech, part encouragement, part exhortation to keep working, then turned them loose. Tom grabbing his bike gear, winking at James and heading for the door. Katy and David already deep in discussion as to how to link their outputs for better analysis as they followed him.

Harding stayed in his seat. 'A word, James, if I might?'

Susan gave James the look. Much the same as the one he'd become accustomed to in the army. And walked out, heading for her next meeting.

'Sir.'

The DDG leaned back in his chair with a yawn. 'You don't need to call me sir, James. We work on a Christian name basis these days.'

James smiled. 'I'm pretty much accustomed to using the term.'

'I can imagine.' Harding looked at James for a moment before speaking again. 'I apologise for that comment about Bale earlier. It was uncalled for, given his – and your own – sacrifice in the line of duty. Even if those waters were then a little muddied by the abrupt nature of his subsequent departure from the Metropolitan Police Service.'

James nodded, acknowledging both Harding's point and his apology.

'He's resourceful enough, if that was what you were asking. If there are secrets to be uncovered in the Salagin household, he's as good as anyone else to sniff them out.'

'But you're worried about him.'

'I would have to admit a certain fondness for the man, if pushed.'

Harding nodded. 'Understandable, given your shared experience.'

'But unwise?'

The DDG looked up at the ceiling for a moment. 'We quickly learn not to have close friends in the Service, James. I've lost more than one agent in the course of an operation like this. And the closer you are to them, the harder it is to cope with their loss. I know...' he raised a hand to forestall any comment '...that's easier said than done in your case. But you'll have to harden yourself, I'm afraid. Bale is an asset. And we've put him on the field of play as a somewhat unwilling member of the team, which means you'll need to keep the pressure on him.'

'And be prepared to lose him?'

'Yes. And after all, you're not just a soldier. You're ex-Regiment. You must have experience of that sort of thing. So treat Bale like you would any soldier. Do your best to keep him operational, but in the event that his potential loss might save hundreds or thousands of lives, do your job and take that risk. That's all I ask.'

22

Mickey presented himself at the address he'd been given. Unsurprised to be greeted by a suited butler. A man with an apparently bulletproof mien of imperturbability. One that Mickey suspected was regularly put to the test by Albie. Battered by the gang leader's blunt take on life. But then money could paper over many cracks, he decided. One glance at the vaulted entrance hall proving his point. Paintings, sculpture and furnishings. All tasteful. All with the look of serious money. And none of it, he guessed, chosen by Albie.

The butler took him through the house. To what Mickey guessed was likely to be Albie's inner sanctum. Where he expected to have his collar fitted, nice and tightly. And the terms of his servitude explained to him. He stopped in the doorway and looked around. A good-sized room, with two entrances. The one he was standing in. And a pair of glazed doors looking out on the garden. The transparent panels a shade of blue that indicated armoured glass. Presumably to discount the possibility of anyone breaking into the house through them. Or shooting Albie through them, for that matter.

'Well now, here he is! Just like I said he'd be, all bright-eyed and bushy-tailed at the thought of working for his uncle Albie!'

Albie, sitting at a desk big enough for a decent game of ping-pong. Wreathed in cigar smoke and smiles. Keith, the minder from the car, sitting on the sofa. Still armed, pistol in a shoulder holster. Still looking hostile.

'I still can't see what all the fuss is about. If this geezer is the one that did for Joe Castagna, then Joe Castagna must have been about as hard as day-old dog shit.'

He got to his feet. Strolled over to where Mickey stood in the doorway. Looking him up and down from a foot away. Deliberately intruding into his personal space with a smile that bespoke his readiness to do violence. Albie pulled at his leash, so to speak.

'Leave it out, Keith! He's on our side now! Ain't yer, Mickey?'

Mickey ignored the goon. And nodded tersely to Albie.

'I'm here as requested, Mr Ward.'

'He's got some spirit, hasn't he?' Albie smirking. 'Requested? Ordered, more like! Come over here, sunshine, and let's see what you've got for your uncle Albie, eh?'

Mickey stepped around the immobile Keith. Close enough to smell the big man's sweat. He approached the desk. Tossed a velvet bag onto the wooden surface in front of the gang leader. Albie opened the drawstring neck, tipping out the contents into a glittering heap.

'Oh yes. This looks very tasty. How much is this lot worth then?'

Mickey shrugged. 'I never got round to having them valued. They were part of Castagna's emergency fund.'

The gangster reached into his desk's top drawer. Pulled out a strangely proportioned revolver. Standard grip and barrel, but the revolving cylinder containing the bullets twice as long as it needed to be. Put it on the wooden surface and went deeper into the drawer. Mickey taking a good look.

'Smith & Wesson Governor. I've never seen one before.'

Albie grinned up at him. 'I bought it for the name. Governor, right? Like me. My watch is a President and my gun's a Governor. And it takes a .410 shotgun round, which makes it proper tasty for a kneecap. Or just for keeping heads down.'

Mickey nodded. Keeping to himself the fact that he'd take a Glock 19 over the deformed wheel gun any day.

'Ah, here it is.' Albie came up with a loupe, putting the pistol back in the drawer. 'Right, let's have a butcher's...'

He held one of the stones up to the light. Had a good hard stare at it through the loupe. Then grinned like the cat that got the cream.

'Nice gear. Mickey son! It's the four c's that matter with diamonds: cut, colour, clarity and carat. Cut, well they're round. Never go wrong with round. Colour, perfectly transparent. Clarity... this one's flawless, so they probably all are. And carat. This one's a nice one carat, perhaps a little bit more. Perfect for a jeweller – they'll buy them all day long when they're this good.' He raised the stone to the light. 'Like currency, this. I wouldn't take less than ten bags for this little beauty.'

Albie looked at the pile of diamonds. Calculating. Then scooped them up and put them back into the bag. Mickey watching, poker-faced.

'That's a million quid's worth right there! I told you this bloke was the one we needed to find, Keith!' He looked up at Mickey. 'All those months you spent in hiding, and you had to blow it by coming back. Good thing you did though, or I couldn't have cashed in on Joe's savings. Or been able to secure your services, eh?'

Mickey nodded impassively.

'You want me to tickle a bit of respect into the cunt?' Keith, his voice loud in Mickey's ear. Mickey restraining the reflex to throw an elbow into his throat and follow up with a roundhouse kick. Do some freelance dental reconfiguration while the goon was still getting over the first shot.

'Nah, leave him alone.' Albie, doing amused but merciful. 'He's just sprung me loose a million quid in sparklers. And now he's going to go and deal with a small but persistent niggle for me. Ain't cha, Mickey?'

'You give me details, I'll make the problem go away.'

Albie leaned back in his chair. Utterly in control. Utterly smug.

'The details are simple enough. The Mantle twins have threatened one of my clients. A meat packer, by trade, name of Gibbins. Told him that he either pays them the insurance money he usually pays to me or bad things will happen. Which shows the sort of balls they've got, walking into a shed full of men with knives and making threats. They need putting down before they become a problem. I'll leave the details to you.'

Mickey shrugged. 'I'll need a weapon.'

Albie laughed knowingly. 'Too fucking right you will! They're tooled up to the eyeballs, the Mantles. Rifles. The kind with the magazine behind the trigger. And even without the firepower they're a pair of right nasty bastards. Go and see my man Errol, he'll sort you out with whatever you need.' He flicked a card across the table. A single phone number written on it. 'Come back when the job's done with some proof that you did it. And then we can talk about your longer-term *employment*, right?'

23

The next day. Mickey's late turn in the Salagin household. One 'til nine. Leaving the morning free. James collected him at nine, and they drove over to see Albie's gun dealer Errol.

Stopping for breakfast on the way. James outlining what he gleaned from the Met's intelligence database with regard to the Mantle twins.

'They seem to think they're some kind of latter-day manifestation of the Kray brothers.'

Mickey raised an eyebrow. 'And what, Colonel Mustard, gives you the first idea what the Krays were like? Having watched Tom Hardy play them both and knowing they were good to their mum doesn't really count.'

James shook his head in mock dismay. 'Oh no, you've seen through my delusions of being a normal person.' He leaned forward. 'The thing is, Mockney Michael, I have been around a fair bit. And met quite a lot of people who I suppose you might call "tasty".'

Mickey pursed his lips. Opened his hands in recognition of impending defeat. 'I take it all back. I had no idea you had a doctorate in twentieth-century gangs of London.'

James ignored the jibe, his voice remaining completely factual in tone. 'I had a sergeant in the Regiment whose father used to do small jobs for them. The Krays, I mean. He was what was termed a "slap man".'

'A slap man.'

'Indeed. When someone did something the twins found disappointing, they would send him round to point out the degree of unhappiness they were feeling. By administering what they called "the slap".'

'And this was fair warning, I presume?'

'You have it. Were the slap to be ignored, the transgressor would be informed and a punishment of greater severity would be visited upon them. Of course more than one slapping turned into a full-scale fight. Not everyone being happy to have the slap delivered. And it was at this point that the fact that my man's father was every bit as much of a psychotic as the twins came in handy, it seems.'

'Interesting. And there's a point to the story, I presume?'

'Oh yes. The Mantle twins, it appears, are often accompanied by three associates who served in the army with them. And who are therefore very familiar with the use of the weapons the twins managed to spirit away when they were discharged. So I suggest that we equip ourselves to cope with any threat that might be posed by these latter-day slap men. Seems only sensible.'

Errol, it turned out, was a higher class of arms dealer than the usual. Operating an outwardly legitimate gun shop in Barnet. Heavy steel shutters. Anti-ram-raid bollards. Panic buttons and state-of-the-art security. His premises tastefully decorated. The weapons all secured in steel cabinets, each one chained to its rack.

Legal firearms and ammunition for the club shooter. Auto-loading twelve bores for the practical shotgun enthusiast. Bolt-action target rifles in a variety of calibres. Expensive optics for the wannabe sniper. State-of-the-art air rifles for the die-hard pellet enthusiast. And a display case of automatic pistols for the section-five firearms user. Artfully displayed with loaded magazines beside them. Dummy rounds, but no less enticing to a potential user.

Mickey and James admired the set-up while Errol dealt with his previous client. Calm, courteous and professional. With the previous client sent happily on his way, Errol turned to them.

'You're Bale?'

Mickey nodded.

'And I'm—'

The gun dealer raised a hand to silence James. His previous courtesy having clearly decided to take the rest of the day off. Making way for hard-eyed suspicion. 'Albie said there'd be *one* of you. And I don't like surprises. So I suggest you go back to Albie and tell him that...'

He frowned as James raised his identity card.

'What, you're filth? I already paid off—'

James smiled. Shaking his head. 'You've paid off the police? We'll add that to the list of problems we have with your business, shall we? I'm not the police. I'm MI5. And we don't jail people like you. We just make them disappear.'

Total nonsense, as James knew all too well. But just the sort of thing that the criminal fraternity loved to fantasise about. That there was an even bigger set of bastards than them running the country. Wanting to believe. Praying that there was someone in authority ruthless enough to defend the realm. Someone as ruthless as they told each other they'd be.

Rather than the borderline white-collar criminal ex-Etonians that was the sad reality.

'We've got hard evidence that weapons you've sold from these premises have ended up in the hands of Islamic fundamentalists.' More falsehood, but James delivered it convincingly enough to make Errol go pale. 'So, without wishing to sound like too much of a cliché, we can do this one of two ways. Pick one.'

Errol raised his hands, palms towards the two men. 'I get immunity from prosecution, I'll tell you whatever you want to know.'

James shook his head. 'I can do better than that. You're going to equip us for the job that Mr Ward wishes Mr Bale here to carry out. And tell us everything you know about the Mantle twins. After which we can have a nice long chat about Mr Ward.'

'And that's all? No prosecution?'

Mickey leaned forward. 'If dealing with you would take some guns off the street, you'd already be blindfolded and in the back of a van.' Allowing the implication of blindfold rather than cuffs to sink in for a moment. 'But the hard truth is that if we nick you someone else will take up the slack. So as long as you behave yourself, and keep your mouth shut with regard to our dealings, there'll be no prosecution for the time being. Think you can keep your end of that bargain?'

Errol nodded briskly.

'Good. In which case we'd better have a look at the *real* stock, hadn't we?'

24

'What do you make of Bale?'

Pavel Salagin looked up from his paper. The sports pages, his usual read in the car. Dissecting the media commentary regarding his F1 team with a keen eye. Having more than once instructed his lawyers on the basis of a few unwary words. Teaching commentators and agents alike to be a little more circumspect around him.

'What do I make of him?'

Kuzmich, half turned in his seat, nodded. 'Your impression, Pavel Ivanovich. After all, it was you who employed him.'

His friend twitched a lip in a ghost of a smile. Both men knowing that Kuzmich's advice had been to let the opportunity pass.

'Michael Bale is...' The oligarch thought for a moment. 'Intriguing. He is clearly capable of great violence. And of taking punishment too. You know he was shot defending his principal.'

Kuzmich snorted. 'Better not to have been shot.'

'But nevertheless, Yuri, he... now what is the parlance the British use...' Salagin thought for a moment. 'Ah yes, he *stopped* a bullet. A rifle bullet, not your everyday nine millimetre. And still he had enough left in him to kill the last threat.'

'He has grit. But you know what they say about something who seems too good to be true. And I think that this ex-policeman is literally too good to be true. If—'

Salagin's phone rang, and he raised a hand to call for silence. 'It is William.'

Kuzmich pulled a face. Salagin's team principal not being very much to his taste either. Too much passive aggression, and without any discernible ability to switch to the full fat version. Salagin listened to what the other man had to say, a slow smile spreading across his face.

'I see. He actually told you that himself? This isn't just his agent flying a kite?'

He listened again, nodded slowly.

'Very well. Yes, we can afford that, if he's making a genuine offer. He could be just what we need. What? No, no texts, no emails. Nothing on the record. I'll call you later on a secure line.'

He put the phone back in his jacket pocket.

'Well that is good news. And here we are.'

The car had pulled in to the usual parking spot outside the building in which GazNeft was based. A spot designated as drop-off and pick-up only. A rule that his drivers routinely ignored. Under orders from Kuzmich to be outside the door when the boss walked out. And to smile and accept the inevitable parking tickets.

The two Russians got out of the car. This being a meeting with the Watchmaker's operational staff, rather than the man himself. And therefore to be attended by Kuzmich as well. Salagin's presence clearly more to ensure he stayed fully implicated than for any practical purpose. They went through the usual routine, badged and took the right-hand elevator. Emerged at the GazNeft reception. Were greeted and ushered through into the closed section of the floor. Separated from the office functions by opaque glass. The Watchmaker's palace of delights.

'Pavel Ivanovich. Yuri Sergeivich. This way please.'

The woman who posed as Chasovshchik's PA was waiting for them. Leading the two men to a meeting room. Three more members of her staff waiting around the table. Greetings were exchanged, coffee poured. After which she gestured to her assistant, who flicked an image up onto the big screen.

York Minster, pictured from the air. Its stone bulk open to the skies for much of its length. The central stone tower having collapsed in on itself. While the heavy timber beams of its roof were missing. Catapulted from their seven-hundred-year resting places onto the surrounding houses. Causing obvious mayhem. A dozen ambulances scattered about the Minster's cloisters. And the same number of fire engines clustered close into the ruined cathedral. Damping down the embers of the fire. A conflagration that had gutted what hadn't been pulverised.

'So, we have inflicted two strikes. The first a nominal failure, although the media impact of the dead policemen was high. And one success. Exactly as planned. And it's clear from the media that the tide is turning our way. Far right groups are growing more vocal by the day. Causing Muslim communities to turn in on themselves. And in turn giving encouragement to the men among them who wish to strike back on their behalf. Their radicalisation reinforced by the presence of ISIS members who infiltrated the UK in the Syrian diaspora. Resulting in several dozen men with a burning desire to martyr themselves for their religion. An idiocy we are happy to exploit.'

'We still have complete visibility of these groups?'

She nodded at Salagin. 'All tidily catalogued for us by MI5. Including one which has been compromised by infiltration. Information laid wide open by our specialised operators in Moscow. Three of their groupings are under close surveillance; another dozen are monitored passively, as long as they refrain from too much talk on Telegram. We have a list of people that we need you to deliver messages to. Messages on paper, not electronic. Recruiting them for the final act of this campaign. And advising them how to prepare for their roles. How to remain under the radar. And what to do when the time comes. We have chosen their roles carefully, given what we, and more importantly the British Security Service, believe their capabilities to be.'

She pushed a sheaf of paper across the desk. Names and addresses on one. Individual messages several paragraphs long on the sheets beneath it. Handwritten to avoid any trace of them on a system. Kuzmich leaned forward and took them. Reading one of the detailed sets of instructions.

'Make sure you get each one to the correct recipient. There are some very explicit instructions for some of them.'

Salagin leaned back in his chair. Not needing to read the instructions to discern their purpose. 'You intend using them for the final spectacular.'

'Yes. It will add a dimension of terror that the inhabitants of this city will not forget for a very long time. And besides, the surprise when they all pop up on MI5's radar at the same moment will be priceless. But that is all for the final weapon's deployment. Before that we have the fourth Sunburn to deploy. To strike a blow that will strike fear into the minds of millions of this country's population. And cost billions of pounds to repair. Yuri Sergeivich, you will take the essential equipment to the *Prizraki?*'

'Yes, as ever.'

The usual secure container was passed across the table. Its bang button masked by a stiff plastic dome. Requiring only a sharp squeeze to snap the cover and actuate the self-destruct. Reducing the contents to deniable fragments. A fingerprint reader the only means of opening it.

'And now I can you what the target is, and what assistance we will need from your organisation. As ever, you provide the shopping; we'll deliver it to where it's needed.'

The assistant flicked up a new slide. An image of the target for the third strike.

'You don't pull your punches, do you?'

The planner looked across the table at Salagin. 'Orders from the top, Pavel Ivanovich. These people are to understand just how it feels to have their country taken to pieces by a ruthless enemy. If the British want to fight, we'll show them what it means to take on an enemy as hardened to destruction and bloodshed as the *Rodina*. But if that impresses you, wait until you see what we've come up with for the last weapon.'

25

L unchtime Sunday. Albie's man Gibbins less than enthusiastic when Mickey rocked up at the meat-packing house. Having, Mickey suspected, already decided to switch allegiances. He looked at Albie's new enforcer with an expression of exasperation.

'I told Albie that he either protected me or I'd have to pay the Mantles instead of him. I thought he'd bottle it, to be honest. But I didn't think he'd send one bloke. One *fucking* bloke? I don't know what you think you're going to achieve on your jack. Have you *seen* the Mantles in action?'

Mickey shrugged. 'Ignorance is bliss. But the question is, who's the ignorant party?'

The pack-house owner threw his hands up in anger. 'Fuck all that clever bollocks! You show out to the Mantles, they're going to know I warned Albie they was coming today! You might as well fuck off quick sharp before they get here. Because if you're still here when they show, you'll get put in a box and they'll have me for squealing. Just tell Albie that he—'

A man in a blood-spattered white coat wearing the long chain mail gloves of an industrial butcher put his head round the door. 'They're here.'

Gibbins shook his head in disbelief. 'Fuck! You can still get out of this in one piece; just use the back door...'

He watched in disbelief as Mickey strolled out into the pack house. Which, Mickey noted, was suddenly and unsurprisingly empty. The workers having clearly decided to take a break. And thereby avoid becoming collateral damage. Five men walked in through the open loading bay door, two out front and three a few paces behind them. While the brothers appeared empty-handed, their hangers-on were seriously tooled. All three carrying L85A3 automatic rifles. The version of the army's assault weapon that actually worked properly. Unlikely that there'd be any jamming.

'Who's this then, Gibbins? You recruited some muscle to see us off?'

The speaker was the slightly taller of the Mantle twins. Darren, if Mickey's research was correct. Having taken the liberty of tapping up an old copper mate for information. Photos. Army records. Intelligence. Darren being the gang's leader, to some degree. The less psychotic of the two brothers. And somewhat more calculating.

His calculations including a knowledge of his brother David's urge to do harm. And the ability to use him like a fire and forget missile. Because David, once unleashed, would stop at nothing to put his man down. And keep him there. Not that Darren was shy when it came to fisticuffs. But knew that sometimes only a proper mad beating would make his point.

'I got you the money, Mr Mantle.'

Gibbins started forward. Froze when Darren raised a hand with the palm out. Taking a good look at Mickey. Clocking his body language, confident. And his hands. Empty.

'Who's this old fart?'

Mickey, hands in his pockets against the pack house's chill, grinned back at him. 'I'm here on behalf of Mr Ward. Either to make peace or to go to war.'

Darren shook his head, looking down at the floor for a moment. 'Oh dear. Looks like it's war then. Because we're not peaceful-minded. Especially not when it comes to Albie Ward.' He skewered Gibbins with a stare. 'The price I named you was for an orderly handover of control. This doesn't look orderly to me. This looks like you went to Albie behind my back. Like you told him what was happening. Doesn't it?'

The butcher spread his hands in protest. 'I didn't have a choice in the matter! I swear I had no idea Albie was going to do anything about it until this muppet turned up!'

Darren nodded slowly. 'I can see how that would happen. But the facts remain the facts. So, either I take David's muzzle off him. Let him give you a good old chewing. Or you go and fetch the extra ten bags that I warned you I'd need if we had to overcome any obstacles to this new arrangement.'

The meat packer gestured to Mickey. Who was standing motionless in the middle of the pack house. Carefully doing nothing to provoke the men with the rifles. And eyeing up the twins.

'Does that look like an obstacle? You'll go through him like shit through a goose!'

Darren shrugged. Taking off his coat. 'Not only does "that" look like an obstacle, but it looks to me as if I'm going to have to remove it. Given I doubt you can.' He waved a dismissive hand. 'Now shoo. And don't show your face without the extra readies.'

He turned to Mickey.

'Nothing personal about this, buddy. If Albie sent a single man to front us up he's more of a mug than I thought. But an example will have to be made, right?'

Mickey nodded equably. 'I can see how you'd view it that way.'

He removed his own jacket, watching as Darren fished in a pocket and came out with a set of iron knuckles. Having left his own in the flat. Knowing just how much their use would limit his fighting style. Darren walked forwards in a business-like manner. Administering a beating that was likely to leave the recipient maimed all in a day's work. Coming straight at Mickey, fists clenched around his knuckledusters.

Mickey had originally intended relaxing in Thailand. Getting some sun. Drinking some beer. Learning to scuba dive, perhaps. All of which had gone out of the window on day three. When he had realised that there was a mixed martial arts gym across the road from his rented flat. Intrigued, he'd strolled down to watch. Surprised to find an MMA school in a country so dedicated to the indigenous martial art – Muay Thai, a ferocious and respected blend of fist and kick-boxing.

He'd leaned against the frame of the open-fronted gym. Watched as the pupils went through their paces. Nodding with approval when they threw good punches. Wincing at the ferocity with which they went at each other. And would have left it there. Gone back to his beer and sunbathing. If the trainer, an American, hadn't beckoned him over at the end of the session.

'Hey. I'm Chuck.'

'Mickey.'

Mickey shaking the offered hand. Feeling the callouses of a martial artist's long practice.

'Well, Mickey, you don't mind me saying so, I think you should have been on the mat rather than watching.'

'How do you make that out?'

'Simple, man. I was watching you, off and on. And you were twitching like a dog watching cars go by and wanting to chase them. You *sooo* wanted to be throwing those punches, yeah?'

Mickey had shrugged. 'Old habits die hard.'

'You're a Brit. So you, what, box a little?'

Mickey had grinned. 'Quite a lot. Enough to not want to get kicked like that.'

Chuck had shrugged. 'Guy your age needs a little caution. But what if I told you I could add twenty pounds of muscle to you? Fighting muscle, not just bulk. Teach you how to deal with the sort of shit you look like you like getting into?'

And with that, Mickey had been hooked. Spending most of the next year training and sparring every day. Gradually muscling up and toughening up. Learning to fight without restrictions. Knowing he'd never be good enough to do it professionally. But conditioning his reflexes to deal with skilled amateurs. And learning a lot about one-punch knockouts.

So when Darren doubled his pace and came in swinging, Mickey was more than ready. Slipped the first punch: a straight right. Close enough to feel the fist's draught on his left ear. Knife-fighting distance. Then stepped back in a classic pull. Putting Darren on his left. Tempting him to extend his scything left hook further than was wise. Parrying the punch with a lateral block. Pushing his opponent further to his left. Then stepped back in. Unveiling his true threat. And firing in a merciless right hook into the side of Darren's body.

Putting the shot in low, below the ribcage. Then stepped back. Knowing the fight was already over.

Darren's eyes narrowed in sheer disbelief that he'd been hit. An expression that lasted for all of a second. Certain, in that blink of time, that he was going to kill this mug for his temerity. And then the liver shot hit him. He staggered. Disbelief replaced by disorientation. Then folded to the floor, unable to stand. His blood pressure suddenly dropping like a stone. Collapsing, his body's automatic reflex to avoid blacking out completely.

'What the fuck did you do?' Mad brother David. His muzzle literally coming off before Mickey's eyes. Staring at his brother in disbelief. 'Did you fucking shank him?'

'He's all right. Here, come on, buddy.'

Mickey took a firm hold of a stunned Darren's right arm. Half-lifted him off the floor and then twisted the limb violently. Putting a thrust kick into the helpless gangster's side. A move Chuck had taught him as a last choice, in a fight.

'*You do this to a guy, you're going to dislocate his arm and wreck his rotator cuff for life. Needs surgery to correct it, and even then it's never quite the same. So last-resort stuff, right?*'

Or for when you really don't want the guy getting up again. As the American hadn't needed to add.

Darren screamed in agony.

And David roared in anger. 'Fucking shoot the cunt!'

A bang split the air. So loud that one of the gang members actually dropped his weapon. All four of them turning to find James in the doorway. Looking them down the iron sights of Errol's twelve-bore autoloader. A bullpup mag holding ten rounds behind the trigger grip. And a heavy barrel with a muzzle like an underground tunnel. A threat so naked that no comment was needed.

James having put the warning shot into the carcass of a cow awaiting butchery. Which had been blown apart by the blunt-ended slug round's colossal impact. The lump of steel then having kept going to impact the far wall. Showering breeze-block shrapnel across the floor. His voice a whip-crack of authority. Parade-ground honed and invested with the authority of superior firepower.

'Any man holding a rifle by three gets the next round! One! Two!' He leaned forward into the weapon. Putting the sights firmly on his next target. 'Thr—' The clatter of the other two weapons hitting the ground told its own story. 'Lie down, hands on the backs of your heads! GET DOWN! NOW!'

The three goons dropped into prone positions. While David turned slowly to face the shotgun's threat. Taking stock of James's stare with the cunning of the partially deranged.

'Nah, you ain't gonna shoot me. I ain't tooled. But I am going to rip his guts out for that...' Turning back to face Mickey. 'So you either shoot me, or I bite your mate's face off!'

James shot a glance at Mickey. Who nodded. Raised his hands. And did the 'come and get it then' hand gesture. Channelling his inner Keanu.

Seven steps, Mickey reckoned, as David started toward him. Showing every sign of being willing to follow up on the threat of facial surgery. Took a step forward himself. Getting his balance right. Pivoted and met the onrushing psychotic with a midriff kick. David slowing his advance, and reaching gleefully to grab his foot. Except it wasn't there. Mickey pulling his foot back and resetting. Dummy sold, target briefly static. David's head now four feet off the ground, rather than six.

The straight thrust kick was delivered with all the strength in a very muscular thigh. Putting his booted heel into David's face. Breaking his jaw and giving him a mouthful of shattered teeth. Mickey stepped back and to the left, then attacked again. While David was still trying to work out what happened. Mickey going in low with a side thrust kick to the thug's right knee that folded his leg sideways. Tearing both ligaments in a moment of fearsome pain. Provoking a gargling scream of agony while the flailing gangster decided whether to swallow his teeth or spit them out. Mickey reset again, ready for a further attack.

'No need, Michael.' James took over. Suggesting to the prone muscle that they take the hapless twins to Accident & Emergency. Covering them with the shotgun while Mickey collected the rifles and put them in the boot of James's car.

'I can't fucking believe it.' Gibbins, staring after the slow retreat from the pack house. His face brightening. 'On the upside, I get to keep the twenty thousand they wanted from me.'

Mickey shook his head. Wondering momentarily how anyone could be so naive.

'You know Albie's going to ask what happened. And you know I'm going to have to tell him everything. Because if he doesn't hear it from me he'll get it from someone else. And then he'll take it out of me. So give it to me now, eh?' Mickey, having a very specific purpose for the money, put his hand out. 'Don't make me take it from you.'

26

Mid-afternoon Monday. The day after the encounter with the Mantles. Mickey was making the most of an opportunity to eat in the UGS kitchen. Dismissed to stand-by after escorting Salagin to his club for an all-morning meeting. More to be seen, he suspected, than for any useful purpose. But if that was what floated the Russian's boat, who was he to argue? Now he was tucked away at the staff table in the corner of the kitchen. And halfway through a plate of Chef's sensational lasagne. Wondering if an hour in the gym after work would compensate for another helping. Inwardly bemoaning the lack of a big jammy glass of Primitivo.

Not that Mickey was any sort of wine expert. As Roz had frequently reminded him. Usually in defence of her favourite sommelier's latest attempt at robbery. Serial wallet-lightening through the medium of Roz's liking for posh wine. Paulo describing the rich Italian red as 'intensely flavoured and deeply coloured, rustic, chewy and with a sweet finish'. Which had all been so much bollocks to Mickey. He just knew what he liked. Not that Primitivo would have been his first choice, given access to Salagin's wine cellar. Mickey just knew that

any cellar with a thousand-bottle rotary dispenser wasn't
going to be doling out the cheap shit.

Kuzmich entered the kitchen at the other end of the room.
Looked around. Found Mickey. Not exactly hiding, but
keeping his head down. And smiled. Crossed the room at an
easy pace. Pulled up a chair and joined Mickey. Not asking
permission, of course. Picked up a fork and carved off a piece
of Mickey's dwindling lasagne. Much to Mickey's poorly
hidden disgust. Chewed enthusiastically, making enjoyment
noises. And eyeing the rest of Mickey's meal. The fork poised
to strike again.

Mickey kept his voice low. Avoiding arousing the interest
of the kitchen staff. But his tone close to that of a junkyard
dog deprived of its bone.

'You put that anywhere near my dinner again and the
plate's going to be in your face. Followed by me.'

. The two men locked eyes. Mickey raising his eyebrows to
send the Russian a very clear message. I *might* not win the
fight. But I *guarantee* you won't enjoy it. Mickey knowing
that he might well be something of a surprise to Kuzmich.
Who didn't look like he'd worked a speed bag any time
recently. The Russian grinned. His teeth still flecked with
meat sauce.

'I like you, Michael Bale. You respond to every little push
in a… how do you say it… *proportionate* manner. You draw a
line and invite me across it at my own risk. Telling me without
saying so that you will resist with your full capability.'

Mickey nodded equably. Forking up the remaining large
mouthful of lasagne. Filling his mouth to give himself an
excuse not to speak. And pondering the inclusion of words

like 'proportionate' and 'capability' in Kuzmich's comment. Notching up his estimate of the Russian's linguistic abilities.

The Russian sucked the fork clean. Putting it back where he'd found it. Another insight, Mickey guessed.

'I have come to tell you that you have a job to do. An important job. Pavel Ivanovich has returned from his meeting early. He needs you to accompany him on a journey to a town in the west of the country. You will be back late this evening. You will drive, of course. And if I know Pavel Ivanovich he will sit up front with you. I counsel you to be careful in conversing with him.'

Mickey swallowed the lasagne. Leaned back and grinned. 'Speak when I'm spoken to? You do know that I've spent more than a decade guarding senior politicians and the royal family, right? Because we pretty soon learn not to ask the boss how it's hanging.' Shut his mouth and waited for the other man to react. Disappointed to see no reaction at all.

'I counsel you to be careful, in conversing with him. He is conducting a matter of the greatest importance, and does not need to be distracted by anything you might have to say.'

Mickey nodded. 'OK, speak when I'm spoken to. And when does Mr Salagin wish to leave?'

Kuzmich looked at his watch. The Royal Oak an affectation Mickey had noticed before. But never commented on. Because why rock the boat? And decided that a spot of boat rocking might just be called for.

'I've always considered a man like you as better suited to one of the more rough-and-tumble watch brands, Mr Kuzmich. A Casio, perhaps? One of the tough carbon fibre ones with sapphire glass.'

The Russian smirked back. Seeing Mickey's game. 'Yes,

perhaps you are right. I should choose a timepiece beloved of the common man. Something lacking in any style or finesse. And cheap, of course. Marking me out as a man without substance.' His eyes pointedly focused on Mickey's wrist. 'Tell me, your Tudor. How much could I obtain this watch for?'

Mickey grinned. *All right sunshine, let's have at it.*

'A Tudor Black Bay? Less than three bags.'

'Bags? This is not punch bags, I presume?'

'Thousands of pounds. Usually, this model of watch costs three thousand pounds. But this one?' He raised his hand to show Kuzmich the watch face. With the red rose of the royal household. 'Hard to say. But Pavel Ivanovich offered me the Richard Mille off his own wrist for it last week. So perhaps *my* Tudor is worth two hundred and fifty thousand pounds.'

Kuzmich guffawed. 'He is a sentimental man. If not always as astute as his reputation would suggest.' He waved a dismissive hand at Mickey. 'Go and do your job, Michael Bale.'

Mickey nodded. Got up and headed for the door. Slowing down as he passed Chef.

'Careful with that one. He just ate with a fork, licked it clean and put it back.'

And left her to chew on Kuzmich's ankle. He stopped on the stairs, unobserved, and tapped a quick message out to 'his parents'.

'What would you like for dinner later? I'm going out west for a while. Won't be back until late. Eight, perhaps nine, so you might need takeaway.'

The code obvious enough. 'Out west' providing James with a rough indication of where Salagin was heading as a compass heading. Likely duration six hours or so. An estimated outward trip of three hours. Enough to give James the notice needed to scramble special forces into the approximate area. Perhaps even a follow team to pick them up on the motorway.

He opened a Scrabble game app. Imaginatively entitled 'Words'. Product of Box IT. The user interface nothing more than a front. Opened the app. Powering up underlying hidden functionality that Kuzmich's tame hacker Dimitri had no chance of finding without a deep code analysis. A nice clear locator signal for Box to track.

In the garage Mickey looked at the choice of vehicles available. Killing time, while he waited for the principal. Strolled down their line in the garage's soft light. The Lamborghini Sián Roadster. Probably two and half million pounds' worth. The Porsche Carrera GT. Old-school, and another couple of million, easily. And the Bugatti Divo. Which, the internet swiftly informed him, would set sir back a comfy five mil. Plus twenty thousand pounds for an oil service. Shook his head and went into the garage office for a chat with the security guard. Wandered out again. Hearing footsteps, he turned to find Salagin approaching. His perpetual half-smile well in evidence.

'Michael. So you got the message from Yuri Sergeivich. Here is a chance for us to have some fun, perhaps? I am going to a very discreet meeting in a location a long way out on the M4 motorway. It is an unexpected opportunity, but I must be very, *very* subtle. Completely under the radar. So, which car would you chose for this mission?'

Mickey pointed to each vehicle in turn.

'Ignoring the Lamborghini, the Carrera GT and the Divo?

Any of which would be like riding a motorcycle naked apart from a neon G-string!' Salagin nodded gravely. Mickey walked back up the line of cars, pointing to each in turn. 'The 911 Turbo S. Yes for speed and relative ability to pass under the radar. But it's Riviera Blue. Sticks out too much. Nice though.' He moved along the line. 'Taycan Turbo S. A subtle colour. Quick and quite discreet. But it takes an age to recharge every two hundred and fifty miles. While the Merc would be perfect, except everyone knows it's your car. Making it a non-starter for anything discreet. But that...' he pointed to the Audi S8 saloon that Salagin was reputed to have bought on a whim and never used '...is perfect. Quick, comfortable, right under the radar. Unless you think we need to be armed, it's spot on.'

Salagin whipped out that grin that he wore whenever his F1 team finished in the points.

'If we needed to be armed we'd be taking more than just you and I, Michael.' The unspoken corollary being that a good deal of firepower was available, if needed. 'The perfect choice. If you collect the keys we can be on our way.'

Mickey produced the Audi's keys from his pocket. 'I took the liberty.'

Salagin raised an eyebrow. 'Well, now we know that you are a gambler. At least in the small things.'

Mickey drove, obviously. Salagin sitting up front, as predicted. Mickey shooting a glance at his wrist. 'A Patek Philippe that's not a Nautilus or an Aquanaut? That's very contrarian of you, Mr Salagin.'

The Russian raised an amused eyebrow. 'This is a form of compliment, I am guessing. That I do not wear the popular model, but focus on the watchmaker's craft instead?'

Mickey nodded. Steering the anonymous Audi onto the Westway. 'I have to give you respect for that.'

Salagin shrugged. 'It is easy to be of a free mind when you have so much money that there is no longer any pressure to show that you are rich. Those cars. The ones you thought would be like riding a motorcycle without clothes? They sell well, and make their makers very happy, because simply getting behind the wheel of one is sending the same message.'

'That you're too rich to care anymore?'

'Exactly so. It is a strange world to inhabit. Intoxicating, to some degree. And yet it separates you from your fellow man as effectively as steel bars.'

Mickey took his chance to slide the Audi out into the outside lane.

'Feel free to tell me an end destination whenever you like, Mr Salagin. At the moment all I know is to hit the M4 Westbound and not stop until you tell me to.'

Salagin raised an eyebrow. 'You'd like to know where we're going? Perhaps you'd like to see a copy of the agenda for when we get there too?' He laughed before Mickey had the time to decide whether to apologise. 'I jest with you. We are heading for a point close to the border with Wales. I will give you more details later, when I know them. For now, it is secret even to me, known only to the man I am going to meet. Call it operational security, perhaps?'

Mickey nodded and turned his attention to the driving. Pretty sure that James would be looking at his tracking screen. And that the expression on his face would not be a happy one.

27

'So Salagin's heading west? What of it?'

James nodded, acknowledging the point. Having just pulled Susan out of a meeting. And wishing he had some less flimsy justification. One of the analysts on duty had called him over the moment the Words app started squawking for attention. He'd watched the location marker progress out of London for a minute or so before going to find her. Knowing he was close to having enough to mobilise any serious assets. Just not close enough.

'Salagin's heading west, with a likely 120-mile radius if Mr Bale's warning proves accurate. He's on his own, with only Michael for security. As if there's something he doesn't want anyone else to see.'

She shook her head. 'So why take Bale with him? If he were going to contact the ghosts surely he'd have taken Kuzmich?'

James nodded again. 'Agreed. But what if he really wants to avoid attracting any attention? What if he dumps Bale and walks the last few hundred metres to a meeting himself?'

'Have Bale follow him.'

'And do what? He's not armed, so he can hardly take on a team of Russian special forces operators single-handed. And by the time he knows their location they could be gone from it five minutes later.'

'And so?'

'We can assume that this whole thing is innocent. And just tell Michael to observe to the degree he can. Or we put a Regiment covert ops team on standby in the area, and vector them in close once the destination becomes apparent.'

She thought for a moment.

'Very well. A four-man team, no more. Low-profile. Vehicle-borne, no aerial assets. And no fire support. I need to be very sure before I ask the army to put any of their big boys' toys into play. Brief them to observe, primarily, but to be ready to take positive action if needed. If you're right, and this is something more than Salagin going to visit some woman, then we'll need them ready to take down the Russians. And now, if you'll excuse me...?'

She turned and went back into her meeting. Leaving James to go back to his desk and get busy on the telephone. Five minutes later, with his immediate requirements now the Regiment's problem, he walked over to the intel desks.

'Where's our boy now?'

The analyst flicked up a tracking screen with an exaggerated hand gesture. 'Slough. Average speed just under eighty, no major congestion en route.'

James calculated. Perhaps a hundred miles to the Severn crossing. And Hereford sixty miles or so distant from the river. But harder roads to make swift progress on. Making it a good thing he'd stretched his authority. Authorising a police escort for them as far as the M50. Knowing it was touch and go even with flashing blues. A toss-up whether his team would be in the area before Salagin got to wherever he was heading.

28

'I find myself in a reflective mood, Michael. Would you indulge me with your opinions?'

Mickey kept his eyes on the motorway traffic. Remaining vigilant as daylight faded and vehicle lights started to come on. 'By all means, Mr Salagin.'

The Russian raised a finger, visible in peripheral vision. 'Please. You must call me Pavel Ivanovich. You understand the use of the patronymic?'

Mickey grinned. 'I had an introduction to the concept from your cultural attaché.'

The Russian laughed softly. 'I see Yuri Sergeivich has already made his usual impression. So, we are Michael and Pavel Ivanovich, yes? The beauty of the patronymic is that it allows you to address me by my name whilst retaining that essential respect that I realise is important in your role.'

Mickey conceded the point with a slight tip of his head. 'As you wish, Pavel Ivanovich.'

They passed a road sign. Chippenham, Bath and Bristol. Mickey hoping that James could work some Special Forces magic in the time remaining.

'Yes. Well here is my problem, Michael. As you know, my

racing team has been quite successful over the years of my ownership.'

And no wonder, thought Mickey. *Given you put the best part of a billion pounds into it.* The sort of money that buys serious advantages. Designers. Engineers. Facilities. Drivers. Allowing the team to climb out of the pack. With the juice to challenge for best of the rest. Never likely to challenge the really big boys, of course. But dealing out surprises and bloody noses to more famous names. And drawing the sponsorship to allow Salagin to rein back his spending. Although not enough to make it break even.

'They finished fourth in the constructor's championship last year. And they're hunting down third place this time round.'

Not adding that they could already be sitting plumb in third place. If they weren't cursed with a lead driver whose time was well and truly up.

Salagin grunted. 'You do follow the sport then.'

No, Mickey thought, *I'm an old-time Williams fan.* Seemingly forever cursed by their being top dogs when he was at a formative age. It ain't quite the same.

'Enough to know that you bought the team at the right time and for the right price. Then spent the money to turn it around. Made them a name to reckon with again.'

There being an additional dimension to that story. Just not one he could mention. That the man sitting next to him was growing tired of F1's incessant financial demands. If the various media commentators were to be believed.

'Yes.' Salagin sounded reflective. 'And for a few years I was able to bask in that reversal of fortunes. But there is always so much expense. Star drivers. Designers who need to be kept from the clutches of competitors. And of course, palms to grease. That is the idiom, is it not?'

'It is. But corruption in motor racing, Pavel Ivanovich? *Surely* not...'

He shot the Russian a glance. Curious to see how his attempt at humour would be received. Met by an amused eyebrow raise.

'I could tell you stories...'

'Best not though. You hardly know me. For all you know, I could be a journalist.'

The smile broadened. 'Show me such a reptile with that badge of honour you wear, and I will show you a pig with wings.'

'So what's the problem? I was reading only recently that the team is expected to become successful enough to wash its own face commercially?'

'Yes. Every year I hear this expectation.' Salagin putting the *yes, but* note into his voice. 'And every year I have to inject another large sum of money. And our current level of performance is unlikely to result in that self-sufficiency. Not this season, at least.'

Mickey shrugged. 'It's a period of transition. You recruited two new drivers over the winter. Perhaps they need time to adjust...'

'Time? Before very long the other "best of the rest" teams will be over the horizon and long gone. I have a driver mismatch. I know it; you know it. One of them is young, highly talented and ferociously ambitious. Too good for the car, really. A good enough driver that unless he sees a major step up within six months, he will tell his agent to start negotiating with Ferrari or Mercedes. And the other...'

He paused. Shaking his head.

'The other was supposed to be the team leader. A former world champion. A man driven to achieve results. As skilled as an engineer at car set-up. With the delicate hands and feet of a virtuoso. Who has in reality squandered at least a half a dozen championship points already this year. Whose super licence is perilously close to a ban. And who, to add insult to injury, has managed to take his own teammate off the track twice in five races. So you see my problem, Michael?'

'Very clearly. There's no easy answer though.'

Salagin nodded. 'I know. Quite the opposite. There is only one answer. And it is very difficult. But with every race I come closer to realising that it is my only option.'

He fell silent, Mickey knowing better than to intrude on a client's thinking time. After a few minutes the next junction was signposted. Chippenham. The Russian stirred himself.

'Turn off here. Head north.'

Mickey knew the area moderately well. There being a royal household close by. And guessed where they were heading.

'Hullavington?'

29

'They've left the motorway. Heading... north.'

James made the connection. Dialled swiftly. Ops room, Bradbury Lines.

'It looks like Hullavington. Possibly to meet a flight. Have your guys vector in to the western perimeter and get the long lenses out.'

The airfield well known to both James and the Regiment. Once a fully fledged RAF base, now home to a one-time electric car factory and an army barracks. He turned back to the analyst. 'We need to check air traffic. Anything that looks like it might be on approach to the airfield.'

She pulled up another window. Swiftly accessing the national air traffic database. Zooming in to a hundred-mile radius.

'Most of the traffic here is commercial stuff. The only real candidates are that one...' She tapped a helicopter icon heading west along the M4. 'That one...' A twin prop aircraft heading south of Birmingham. Its flight plan popping up as she tapped it in turn. Destination Southampton. 'And that one.'

Another twin. An executive jet, heading north. Flight track

reaching back into southern Europe. Destination Manchester. So nothing with a declared destination that matched Salagin's apparent objective.

'Can you put up potential ETAs overhead Hullavington?'

'Already doing it.'

James stared at the results. Ten minutes for the chopper, but no sign of a turn north to the airfield. He decided to ignore it. Twelve minutes for the closer of the two planes. Twenty for the other. Both of them passing within an easy diversion distance. He dialled the ops room again.

'You've got ten minutes, worst case. Twenty at best.'

The man at the other end took a moment to respond. 'Tracking says the team are likely to be on site and active in... seventeen minutes. Expediting.'

James knowing that there was nothing he could say that would make the inbound team move any faster. The 'Expedite' command enough to have their own blue lights flashing and throttle pinned to the floor.

They watched as the helicopter proceeded on along the motorway. James scratching it from his list with an inner sign of relief. And turning his attention to the two planes. The first still cruising at fifteen thousand feet. The second a few thousand feet lower than before. And descending.

'It's that one.'

Even as he said the words the detail in the flight plan window blinked from blue to red.

'The pilot's declared a medical emergency. Diverting to Hullavington and requesting emergency services.'

James leaned back with a smug smile. 'Now that's convenient.'

30

'You're clean?'

Kuzmich nodded curtly. Waved a hand at the black Land Rover Discovery hidden in the house's unlit shadow. The car picked up from a street in West London. After taking a taxi to a well-known and highly expensive Kensington gentlemen's club. Into which he had walked through the front entrance. And out of which he had walked through the staff door at the rear a minute later. Got into the waiting Land Rover and headed west.

A long stint on the motorway, ten minutes of A road, then off into the back country. Culminating in a slow and careful mile up a rough track. Using night-vision goggles rather than headlights. And stopping to check for any sign of surveillance. The moonless night revealing a blaze of stars. Unhindered by light pollution other than the western horizon's dim sodium glow. The nearest town fifteen miles distant.

'The car was swept for tracking devices before I left London. And I did a scan of my own at a service station an hour ago. My phone is in a strip club in London, which means that anyone who might be interested will believe I'm paying for sex for the next few hours. So don't worry, *Kapitán*, I'm clean.'

Ivan nodded. Stepped back and gestured for his countryman to enter. Kuzmich walked into the house, noting the loaded assault rifle hidden behind the open front door.

'Your security is still tight?'

The Spetsnaz officer nodded. 'Water-tight. We patrol in the daytime, mainly to break the monotony and stay sharp. Nobody ever comes up here, and if they did all they would see would be a rundown cottage. Come on.'

He picked up the rifle and led Kuzmich into the cramped living room. Threadbare carpet, a battered three-piece and an elderly cathode-ray TV. The door on its far side led into the kitchen. And the entrance to another world. The steps down into the underground corridor below occupying the space between worktops and oven. Kuzmich stepped down into the subterranean stairway, Ivan pulling down the hatch that doubled as the kitchen floor as he followed.

The two men walked down the gentle incline of a hundred-metre-long corridor. Wide enough for three men to walk abreast. Or two men carrying a Sunburn warhead. Twelve feet underground by the time they reached the far end. The door that stood in their way was massive. Heavy-gauge steel, digital combination lock. Ivan keyed in the code and heaved it open. Allowing Kuzmich to pass though into the hide as he muscled it shut.

The hide. Three shipping containers buried side by side under several feet of soil. Sleeping quarters, cooking facilities, a dormitory, a gym and a bathroom. Plus storage cradles for six bulky missile warheads. Three of them now empty. With air vents and fans suitably concealed in the house. Whose power kept the lights on and the air breathable.

Constructed at the behest of the newly powerful GRU years before, using money channelled through a web of shell companies. By a builder whose loyalty to Moscow was unimpeachable. The son of an illegal who'd been inserted into his new life fifty years before. A realist who knew that betraying its location would see himself and his aged parents jailed. And who had therefore been willing to cooperate. And to use immigrant labour whose discretion was easily purchased. Or extorted.

The other three ghosts were gathered around the living area's dining table. Looking like the experienced operators they were. Their stares not unfriendly, just questioning.

'Gentlemen.' Kuzmich held up the bottle in his hand. A good-quality single malt. Ivan, gesturing to the spare chair, produced five glasses. The ghosts watched as he broke the seal, waiting until he'd taken the first sip before following suit.

He raised his glass. 'To going home, job done.'

They raised their glasses and drank, but stayed silent. Waiting to hear what he had to say. Knowing that he only ever came when there were instructions to deliver. And the tools with which to enact them.

'The first two targets worked well.' Kuzmich waved a hand to recognise that the first strike had not been as expected. 'Of course the first warhead was supposed to be used in London, but the effect was almost as good. Spectacular, and with casualties. Not just the police either, but the delivery team martyred themselves. Encouraging those others eager to earn their place in the garden of the virgins. And placing every Muslim in the country under suspicion, as far as this country's fascists are concerned.'

He smirked at them, sharing his amusement with the gullibility of the men they were manipulating.

'Which made the Directorate's choice of the cathedral for the second attack masterful. MI5 is now completely focused on the various jihadist groups, because that's where the politicians are forcing them to concentrate. They cannot see beyond the ends of their own noses. And now we have new orders from Moscow. Orders to close this campaign out and withdraw undetected.'

The men around the table went still. Waiting to learn their fate.

'The decision has been made to go for maximum impact with the last two warheads. Which means three things.'

He raised a finger.

'Firstly, we must inflict damage to something meaningful. Something that will be impossible to return to its previous state. Part of the apparatus that makes this country work.'

A second finger.

'We need to inflict casualties. I know none of you enjoys the thought of killing civilians, but it is essential if we are to maximise the power of our strike. The West does not see human lives in the same way we do. For them every life is precious, and the avoidance of death is of paramount importance. So the more people we kill, the better.'

He looked around the table, seeing nothing in their faces other than understanding and acceptance.

'And thirdly, the strikes need to be as high-profile as possible. Media-friendly. We have relearned this from the coverage of the cathedral strike. The screen hours devoted to a single meaningless clergyman have been educational. We need to provide the media with the sensational story that they lust after. And, for the third warhead, the strike will be made at prime time. Piped into every house in the country. Replayed around the world. The proof of this country's decay, that they allow their domestic terrorists to wreak such havoc.'

The sniper, Sasha, leaned forward slightly. 'You want us to film the attacks?'

'Not as such. But Moscow has come up with an idea that will enable us to provide that coverage. You have the target files?'

Ivan got up and went over to the locked steel cupboard in the corner of the living quarters. Opened it and brought out the tablet on which the target list was stored. Handed it to Kuzmich and retook his seat. Who looked across the table at Anton.

'You have three warheads remaining. They are all functional, I presume?'

The armourer nodded.

'Good. Then for the third strike, Moscow has selected target nineteen. It is felt to present the perfect combination of damage and body count.'

He placed the secure transit box he'd been given in the GazNeft office on the table. Applied his thumb to the print reader. Then opened it, and removed the smaller box inside. An input socket for power at one end. And an output socket at the other, for the command wire that would detonate the weapon it was attached to. It excited no comment, the ghosts being used to his delivering the detonator with each new target.

Ivan flicked to the appropriate page of the target list. Turned the tablet to show it to the men around the table. Kuzmich saw eyebrows rise as each man realised what their next objective was.

'Target nineteen has also been selected to enable you to make a safe exit prior to detonation. And to have the maximum probability of avoiding capture. Facilitated, of course, by the resources we can provide you with. The last target is deemed to be of paramount importance. It *must* be serviced, and so you must avoid capture.'

'And the final target is…?'

Kuzmich nodded at Ivan.

'Moscow has decided to make an addition to the target list. Something the mission planning staff came up with recently.' He took a memory stick from the box in front of him and held out a hand to Ivan. Who handed him the tablet. All four men watching as he plugged the tiny drive into its USB socket. Downloading the contents in the blink of an icon and then reopening the list.

'You had fifty potential targets, now you have fifty-one. Gentlemen, I present you with Target Zero. And trust me, if we manage to hit *this* target in the way that Moscow envisages, our names will echo through history. We will be the men who brought a country to its knees.'

31

'**D**o you hear that?'

Salagin had got out of the car as soon as Mickey had stopped it on the appointed spot. And was now leaning against it smoking. Mickey standing close enough to talk. But distant enough to avoid a passive lungful. They had driven onto the airfield easily enough. Their arrival clearly expected by the lone security man. Who had directed them to a parking spot adjacent to the darkened control tower. In the distance Mickey could hear the sound of an aircraft. A thin, mosquito whisper of distant jet engines that was steadily growing in strength. As if in response the runway lights came on.

'This place doesn't look like it's open for business.'

The Russian winked at him. Tossing his cigarette away onto the expanse of concrete. 'It isn't. But then you will know as well as I do that the secret in this life is *who* you know.'

Mickey nodded. Guessing that he had a pretty clear idea of who it was that Salagin knew. And wondering who – or what – might be on board the plane whose navigation lights were descending towards them.

32

James's phone rang. Hereford, sounding quietly smug.
'Sabre is in position. Patching video feed through to you now.'

An icon popped up on the analyst's screen. She flicked it across to a second monitor and then splayed her fingers to open it. The image was muted, washed-out greens. State-of-the-art night-vision equipment. Zoomed in tight on Mickey and Salagin standing by the car. Salagin smoking, and from the look of it talking to Mickey.

'That's our boy.' James spoke into his phone. 'Patch me though to Sabre please.'

He waited. A whisper in his ear announcing the successful patch through. Field noise discipline.

'Sabre to Liaison, reporting as requested.'

'Liaison to Sabre. We have your video feed on screen. Sitrep please.'

'Sabre, jet aircraft on short finals from the north-east. Will report landing.'

James and the analyst waited for a moment or so. Watching the video screen as it tracked the incoming plane down onto the runway.

'Sabre, aircraft has landed and is taxiing to the control tower.' Another pause. While the pilot parked his aircraft. Helpfully pivoting the plane to point back at the runway. Giving the surveillance team a perfect view of its nose. 'Sabre, aircraft is stopped, engines powering down.'

The operator zoomed in, waiting to see which door would open. Then pushed in closer as the port side hatch swung wide. Telescopic stairs opening out to meet the tarmac.

'Here we go...' A moment of hushed silence. And then the team leader's voice changed. Becoming oddly formal and yet, at the same time slightly gleeful. 'Sabre to Liaison... are you seeing what I'm seeing?'

James stared at the screen. Baffled as to what had caused the surveillance operator's apparent amusement.

'Liaison, roger your last, video is clear. One male has debussed from the aircraft and is making his way toward...'

He frowned, falling silent. As the slowly zooming camera managed to frame the new arrival, Salagin and Mickey. The Russian stepping forward to shake the newcomer's hand. With a look that was almost reverential. Mickey playing the part of the dutiful bodyguard, his expression unreadable. Opening the car door for him before walking round to do the same for Salagin. Then watching as the two men shut themselves into the car's spacious rear compartment. With a slightly awed look on his face?

33

Mickey and Salagin got back to the house just after ten. The Russian still furious. Incensed that his supposedly clandestine meeting was now public knowledge. And that a potential driver vacancy would be dissected on the back pages the next morning. One the media hadn't even suspected existed until three hours before. He had tersely bidden Mickey a good night. And then stalked off, in search of a large glass of vodka, Mickey suspected.

'What's wrong with the boss?'

He turned from signing the car back in to find Angela. Doing her evening rounds, he guessed. And looking at him speculatively. Also looking, Mickey would have been forced to admit if asked, very tasty.

'He seems to have been caught in the act of planning to replace Kurt mid-season. With Nathan. That's who we went to meet, and it looks like there was a long lens at the party.'

Nathan. The man who had twice stolen world championships off Kurt's toes at the last race of the season. A combination of outrageous ability, ice-cold nerves and an undimmed ambition to be the greatest ever. But at a difficult age. And having left the championship leaders as the result of their desire to bring in a hot-shoe kid. Cheaper, and the future.

His unexpected ejection from the summit having forced Nathan to scrabble for a seat in the lower reaches of sport with the pay-drivers and journeymen. His huge reputation having secured him a drive with Racing Supremacy. Racing alongside the team's owner's son. Which more than one driver in recent F1 history could have told him was unlikely to end well. Hence, probably, his urge to escape only a few races into the season.

'But how could anyone have found out...?'

'No-one knows.' Mickey lying with aplomb. 'But the odds on Nathan joining Team Rossiya went into free fall. Betting suspended, due to a sudden flood of money. All on Nathan joining the team. Based on presumed inside knowledge.'

She put her hand to her mouth. 'Oh my God! How embarrassing for Nathan!'

'How embarrassing for Kurt, more like. Nathan's probably loving it.'

Kurt being Salagin's current number-one driver. Poached from the championship winners at the end of the previous season. Unveiled with great fanfare at the factory. To all-round celebration by the collective fan base of both team and driver. A match made in heaven, the majority of the media had predicted. With only a few dissenters opining that perhaps Kurt had seen his best years. Everyone hoping for a second coming of the Talented One. The resurrection of a double world champion to be the best once again.

Instead of which Kurt had managed to biff the Rossiya car into the Armco twice. Had beached himself in the gravel at Spa. And had taken his frustrated teammate off the track with him twice. All in five races. Which was bad enough in itself. But made much worse by the additional sin of a blatant refusal to obey team orders. Declining to let his faster English teammate through. At, of all places, the British Grand Prix. Resulting in a hundred thousand Brits baying for his blood. Incensed when an infuriated George had made a doomed attempt to pass Kurt. And found himself shunted off to the scene of a double Team Rossiya Did Not Finish. Bad enough in itself. But as a replay of an incident between the two in the Middle East, when there had been no team orders? Unforgivable. Kurt was clearly on the skids in more ways than one.

Angela leaned in and whispered conspiratorially. 'Perhaps that's where Mr Kuzmich went so secretly, once you and the boss were on the road. To give Kurt the bad news.'

She gave Mickey the 'what do you think of that then?' look. Mickey shrugging. Making a mental note of Kuzmich's absence.

'He probably knocked off to get fed. After all he was nicking my food earlier.'

Angela smirked. 'Or more likely he just knocked off to get knocked off, eh?'

Mickey raised an eyebrow. Put his 'surely not' face on. 'What? You don't think...'

'Well you lot all do it, don't you? After all, I'm kind of hoping *you* do.'

Mickey took a moment to let that sink in. Rolled it around in his mind a bit before answering. 'Well that's very forward of you, Ms...'

'Stewart. Used to be Nolan, but he went over the side on me. So I've been saving myself up for someone with muscles *and* a brain.' She looked him up and down with a speculative expression. The tip of her tongue between her lips, as if she were concentrating. 'I can see the muscles, but have you got a brain, Mr Bale?'

Mickey needed another moment to process that. With the feeling this was all moving a bit faster than his maximum calibrated acceleration. 'I'll let you be the judge of that.'

And felt pretty smug with a pretty much real-time riposte. Until, that was, Angela drove a coach and horses through his repartee. Going, it seemed, for the kill.

'Which is as fine an invitation to a lovely dinner the day after tomorrow as I've had for an age. I'd love to. I'm on late the day after, which will give us plenty of time to get to know each other. All night, if you play your cards right.'

Fuck, she'd done it again. It was like trying to box smoke. Faced with the inevitable, Mickey inclined his head in recognition of his utter capitulation. 'It would be my pleasure? Would madam like to be picked up?'

Angela grinned, delivering the coup de grâce. Handed him a card with her number on. 'Too late, Mr Bale. I've already done the picking up. You can text me the venue and I'll be there at seven thirty sharp, with a thirst. And if you need to take anything to ensure peak performance, get it down your neck beforehand. I'm not one for idle chat while the little blue pill works its magic.'

She walked away without looking back. Leaving Mickey wondering, among other things, just where it was that Kuzmich might have gone.

34

James and Mickey made their rendezvous the next day, at the end of Mickey's early shift. Two teas in a back-street caff. The venue chosen at random by Mickey to avoid the risk of predictability. He grinned as James slumped wearily into his chair.

'You look a bit weary, James. I presume your supervising officer wasn't best pleased last night. A little peeved even?'

'Susan? Peeved?' James raised an eyebrow. Unamused. 'She was absolutely furious. And not just because I persuaded her to put a Regiment surveillance team into the middle of a completely unrelated matter. We could have blown your mission wide open, and she knows it. It didn't help that the bookmakers were forced to stop taking bets on the chances of that man taking over as number-one driver in Salagin's team a few hours later. Every man jack in the Regiment must have tried to get a piece of the action. And since the video feed was going into an ops room with several men in it, it was never going to stay a secret.'

He took a sip of the tea, grimacing at its industrial strength.

'The funny people I work have been emailing me screenshots of F1 gossip sites all day. Apparently the rumour of his impending team change came from "unattributable sources".'

Mickey nodded. His face reflecting equal chagrin.

'It wasn't all that funny for me either. Salagin was like a dog with two dicks after the meeting. Really happy, really chatty. And who can blame the guy? What a coup that would have been, given Nathan's still probably the world's best driver. And then he got a call from Nathan's current team while we were still an hour away from London. Apparently the odds of him moving to Rossiya had fallen from twenties to evens in less than an hour. So they called Salagin and warned him to butt out. And then put out a *very* pointed press release saying he isn't going to be released any time soon. Apparently his contract's littered with penalty clauses to protect them from a mid-season loss.'

He took a sip of his tea. Shaking his head at the memory of the Russian's fury.

'Salagin knew he was beaten as soon as he got the call. He could probably have tickled that trout into his keep net, if he'd had the time to get his ducks lined up. Greased the right palms and got the timing right. But it was never going to work if it came as a surprise to Nathan's current employer. Which meant all Nathan could do was to tell the press he's happy where he is. Nothing to see here. The deal was off before his plane had even got back to wherever it was he flew from.'

'But you're above suspicion?'

Mickey nodded. 'I fronted up straight away. Unlocked my phone and handed it to Salagin. Suggested that he get it scanned by his tame hacker. He told me there was no need, and that he trusts me implicitly. And besides, I had no idea who he was going to meet.'

He leaned closer, lowering his voice. 'My sources in the house tell me that Yuri Kuzmich was on the move last night as well. He left the house about five minutes after we did. And he didn't get back until after midnight.'

'That is interesting. And raises the possibility that the Nathan thing was all just a ruse to take MI5's eye off the ball.'

Mickey reflected on that. Having come to the same hypothesis. Not really liking what it implied. 'I'm playing it straight. Asking no questions, none of my business. Just doing my job.'

'Hmmm.' James shot a calculating glance at Mickey. 'And you're still going through with tonight's meeting with Ward?'

'Doesn't feel like I have much choice.'

'Leave the phone open. I'll be listening from not very far away. If it starts to go off the rails I'll do something.'

Mickey raised a sardonic eyebrow. 'Knock on the door and say how very dare you?'

'Call the police, most likely. Tell them I heard shots being fired. At least that way they'll turn up properly equipped.'

'It won't come to that. He'll be as happy as a sandboy until I show him what's in the bag.'

35

'You're sure you can get it on quickly enough?'

Mickey nodded. Having practised at James's insistence enough times to have the muscle memory pretty much mastered. Made sure the cash was hiding his holdall's other contents. He got out of the car and walked around the corner. Into the field of view of Albie's CCTV. Knowing he was on camera. Walking like he knew and didn't care. Walked down the drive. The front door opening while he was still ten metres from it. Raised an eyebrow at Keith's attempted alpha male stare. The big minder shaking his head. Trying to look forbidding.

'You might fool him, but you don't fool me.'

He leaned forward and tapped Mickey on the chest. Tempting Mickey to break the wrist it was attached to. And then take things from there. A temptation he knew he had to resist. For a little while at least.

'I don't fool *you*? What part of me being blackmailed to do the Mantle twins is it that you don't get?'

'You ain't straight, that's all.' Keith, shaking his head. 'There's just *something*...'

Mickey grinned. Put the holdall on the hall floor and

unzipped it. Then pulled the shopping bag inside it wide for the minder to look into. Since he was pretty sure he was going to insist on examining it anyway.

'Twenty thousand pounds in used notes. Sorry they're loose, but Gibbins wasn't expecting to have to hand over the additional. And there's no way I was going to count money I'm not getting to keep.'

Keith stared at the money in disgust. Knowing that it would be him who had to count and bundle it, most likely. Mickey held his arms out in the 'ready to be frisked' position. Drawing the big man's attention from the money bag.

'Go on, since you don't trust me. Make sure I'm not carrying, eh?'

He stood patiently and uncomplaining as the big man patted him down. Then walked through the house to Albie's office. Keith a pace behind him. Albie was at the desk. His Havana wreathing the room in blue-grey smoke. Keith walked round the big desk to stand behind his boss. Ostentatiously opening his jacket to show Mickey the butt of his automatic pistol. But where Keith was watchful, Albie was ecstatic.

'Mickey, son! You did exactly what was needed! My spies tell me that neither of those mugs will be in any condition to fight back for months. Which means they ain't gonna last a fucking week!' He pointed at the holdall. 'Keith didn't think you'd bring me the additional you took off Gibbins. He's what you might call a bit of a sceptic, our Keith! Whereas I, Mickey, am a firm believer in human nature! And that you ain't stupid enough to try to rob me of a measly ten thousand. Let's have it then.'

Mickey put the holdall on the floor. Squatted down and lifted out the Tesco bag full of cash.

'You *know* I'm not stupid enough to try to have you over.' He put a hand into the bag, coming out with a handful of twenties. 'It'll need counting though. I've only got Gibbins' word it's all there. He put the notes back in, then tossed it across the table. A nice high arc, maximising the loot's hang time. 'Here you go!'

Albie and Keith's eyes following the plastic bag's flight. Mickey depending on that reflex. Wanting all eyes on the prize as he went back into the holdall. Quickly squatting down. Putting his hand into a tight elastic interior pocket and pulling out the cylinder waiting within. Its pin already removed, once it was trapped by elasticated fabric. The lever held in place only by its snug confinement. He tossed it high into the air towards Keith.

'*Catch!*'

And then turned away, pulling on the gas mask he'd grabbed out of the holdall with his other hand. While Keith obeyed the reaction drummed into him from childhood. His hands twitching towards the grenade even as he knew he should be going for the pistol. The tumbling cylinder looking like an easy catch until the lever flipped free as it left Mickey's hand. Which confused Keith sufficiently that he fumbled the canister and let it fall to the floor. Where it lay for a second before vomiting a thick white jet of gas that instantly enveloped the two men.

Keith staggered out of the gas towards Mickey. The automatic drawn, blindly groping in front of him with his other hand. Eyes streaming. Barely able to breathe. Totally unable to see. Making the restriction of Mickey's vision by the mask seem like Panavision by comparison. Mickey ducked away from the pistol's muzzle. Wincing as Keith blindly emptied the magazine in a swift volley. Two rounds clipping the butler as he hurried into the room to see what the commotion was all about. Both bullets blowing a pink mist out of his back. Lights out for Jeeves, Mickey reckoned. Evading the enforcer's questing left hand, he took a firm grip of the pistol. Grabbed his arm with his other hand. Breaking the wrist with a single swift twist. Keith grunting at the sudden pain. The pistol falling from his grip. Mickey pushed the struggling gangster away and kicked him in the balls. Hard. Making sure he wouldn't attempt to recover and reload the weapon with his good hand. Keith hunching over his abused testicles with a squeal of choking pain.

Mickey walked into the gas cloud. Finding Albie on his knees, gasping for air. The gangster looked up at him, barely able to see. Completely unable to speak. Mickey's words were muffled by the mask. Although to be fair he was mainly speaking for his own satisfaction.

'Nothing personal, Albie. Just the way things are, right?'

He dropped the gangster with a straight thrust side kick to the face. Bouncing him off the wall and leaving him sprawled across the floor. Checked the butler, finding him, as expected, already dead. Pure blind bad luck. Nothing to be done there. He went to the garden door and unlocked it. Outside, James was wearing an identical mask.

The two men worked quickly. Using zip ties to truss the gangsters and render them helpless. They secured Albie's arms and legs to his chair, six feet from the desk. Then gagged and blindfolded him. Adding earmuffs playing white noise to complete the isolation.

Then laid Keith across the desk's massive expanse, face down. His arms tied together at the wrists, behind his back.

Feet, ankles and knees zip-tied to render him immobile. And subjected him to similar sensory deprivation. Then Mickey went to make them both a cup of tea. Searching Albie's old-fashioned country-style kitchen for mugs and bags.

'Biscuit?'

'Don't mind if I do, just this once.' James took two. 'All this tying up of semi-conscious men is quite fatiguing. So, what do you think Mr Ward's response to finding himself trussed up like the last turkey on the shelf is going to be?'

'The usual, I'd expect. He's had the sort of power that a man in his position enjoys for too long to believe it could ever happen to him.'

'So what do you think his reaction will be? Death threats?'

'Either that or offers of extravagant wealth. Without any mention of the inevitable death sentence once he's freed, of course. Right, let's finish the job, shall we?'

They returned to the office, taking their teas with them. Mickey carrying a plate covered with a cloche he'd found in a cupboard. Useful for a reveal he was planning. The gas having settled out of the air. Now no more than a fine powder across the office's surfaces. And Mickey got busy opening the safe, while James conducted a somewhat more delicate piece of work. Delicacy being the watchword when it came to the sort of toys he'd brought to the party.

'You're sure that'll work? I don't fancy having an angry Albie coming at me with that oversized pistol.'

James raising an eyebrow at the question. Finishing up his intricate task. 'It'll work. Let's have a chat with Mr Ward, shall we?'

Albie jerked as the headphones came off. And glared as the blindfold and gag were removed. '*Jesus*... my fucking head...'

He looked up at Mickey, then frowned at James. 'Who the fuck is he?'

'Just a friend.'

The gangster's head turned back to face Mickey. Slowly. Like the turret of a battleship. Playing a basilisk stare on him. And Mickey knew that whatever words might come out of his mouth, that was the real Albie Ward, right there. Raging at his humiliation. His voice was hoarse, but harsh with it.

'I should have you skinned alive, by rights. I should make you watch every member of your family die. Slowly and in agony.'

His voice hoarse, from the CS gas. But strong. Mickey had to hand it to him. When it came to whistling in the dark, Albie was quality. Mickey shook his head. Albie seemingly skirting around pronouncing his own death sentence. 'I should' rather than 'I will'.

'Perhaps you should. But it seems obvious that you can't, doesn't it? And I did warn you there were bigger bastards than you in play, didn't I? Want to know who's bigger than you?'

Albie stayed silent. Perhaps already regretting his outburst. And its implications for his own survival. Mickey persevered.

'MI5 is bigger than you, Albie.'

The gangster shook his head in disbelief. 'MI5 my arse! That's spy film bollocks.'

'Suit yourself. Who else do you reckon has access to the sort of weapon that we just used on you pair of mugs?'

Silence. Albie working out his options. 'You want something. Don't you? Or why else wouldn't I already be dead?'

Mickey grinned despite himself. 'There you go. Judging everyone by your own standards, eh Albie? Not everyone wants to kill anyone who doesn't just roll over for them. And thing is, I'm not sure you've got very much left to deal with.'

He lifted the cloche, revealing what was on the plate beneath it. Diamonds.

'What? How...'

Mickey showed him the black velvet bag. Another triumph from Box IT.

'Just reclaiming my property. It was a bit schoolboy of you leaving them in the bag, really. But then all that spy stuff's just bollocks, isn't it?'

'The bag...'

'Yeah. It has some very clever electronics sewn into the lining. And an aerial, to gather and transmit signals. So when you opened the safe, it listened to your unlock code and squirted it to my phone. Of course I didn't just take the diamonds. The remaining contents of your safe are now mine. And the twenty grand. Which will teach you not to make people do what they don't want to do.'

He looked down at the helpless gangster.

'And now you have a choice. You can either rant and rave about how you're going to kill me, or we can put this behind us. You tried to mug me off; you failed. If you try to repeat the trick I'll be forced to do to you what I did to the Mantles. Only worse. So choose. I can either leave you tied up and set fire to the place, or I can free you and we both move on.'

Knowing that Albie was about as likely to move on as he was to shit gold nuggets. The only question being whether he would seek his revenge immediately or with a little more subtlety. Mickey's mental bet being far from anything even remotely subtle. He waited, while the gang leader digested his offer.

'You'll free me, if I promise to let this slide?'

'Yeah. And I'll let you trying to make me your enforcer slide too. All square, peace in our time. Right?'

He saw the cogs turning. The less than intricate reasoning of a man whose one and only driver was dominance. Saw a slow smile slide onto Albie's face. And knew exactly what he was going to do. Shrugged, internally.

'All right. I accept. You let me out of this chair, we can sit down and work out a peace plan.'

Mickey smiled. Putting a mask of relief on this face. Walked around the desk to where Albie sat, helpless. Raised a pair of plastic snips. 'You sure about this? I wouldn't want you on my conscience if you were just stringing me along.'

Albie grinned. 'Nah. I know when I've met my match. You cut me loose, then I'm gonna go and get a drink to wash the taste of that fucking gas out of my mouth. Then we can talk.'

'All right.'

Mickey cut the flexicuff on his adversary's right wrist. Handed him the snips and walked away. 'You get yourself free; I'm going to make a cup of tea. Want one?'

'Yeah, son.' Albie's face so far from unreadable as to be almost comic. 'Two sugars.'

Mickey paused in the doorway. Watching Albie cut the ties on his other wrist. And bend to attend to his legs. James walking briskly past him and down the passage. Mickey wondering if Albie would cut Keith free before reaching for the big chrome pistol in his top drawer. Bad news for Keith if he didn't. Shrugged and walked away down the corridor. Mentally picturing Albie. Cutting away the ties securing his legs. Getting up. Rubbing at sore wrists for a moment and then setting his face hard. Ready to do what had to be done.

Opening the desk's top drawer and reaching for the pistol. With every intention of assuaging that burning anger by emptying six .410 shotgun cartridges into Mickey and James at close range. For starters, at least.

They had discussed the question of Albie at some length. Mickey pretty sure he wouldn't take well to being turned over. Even if it was just payback for forcing Mickey to do his dirty work. His expectation being that Albie would seek a biblical level of revenge. Both men knowing Albie had to be dealt with. If only to avoid him messing up their operation. Neither man really having the stomach to kill him in cold blood. But neither of them in any way deluded as to his likely murderous rage.

Something of a quandary, as James had reflected over weapons-grade English Breakfast tea. And so, after much consideration, they had come up with a fair way to decide the question. By contracting the decision out to Albie himself.

Tea drunk and biscuits munched, Mickey opened the safe. While James had rigged up a field-expedient booby trap. Their chosen means of finding out whether Albie was to be trusted. And to deliver the appropriate punishment if he proved lacking. Using fast-setting epoxy and an assault grenade taken from Errol's illegal weapons cache.

Sticking the powerful bomb to the back of Albie's top desk drawer. Then connecting its cotter pin to the Governor's trigger guard with near-invisible fishing line. And then, with infinite care, he'd straightened the pin's two halves. And eased it so far free of the fusing mechanism that one last gentle pull was all that would be needed to cap off the grenade.

Meaning that if Albie grabbed for the weapon, he'd be his own judge, jury and executioner. Pulling the pin. Releasing the spring-loaded lever. Which would pierce and fire the grenade's first igniter. Whose function was to light the grenade's fuse. Which would then usually burn for four seconds before igniting the second primer. The one that would make the detonator go bang. Usually.

In the case of Albie's grenade, however, the fuse had been removed. Leaving a hollow tube with the second primer at its far end. Meaning that the first primer's ignition would light up the second primer milliseconds later. Firing the detonator into two hundred grams of explosive an instant after that. The results likely to be catastrophic. Given James had chosen an assault-grenade from Errol's trove of delights.

Assault grenades being designed for fighting at close quarters. Lacking the notched wire casing used in defensive fragmentation bombs. Used to shower a ten-metre radius with shrapnel. In its absence, the grenade relying on simple brute explosive power. The aim being that a well-thrown or projected grenade would land close to an enemy combatant. And kill him instantly, by simple explosive overpressure. But without any threat to the thrower. A small kill radius, perhaps two metres. But inside that distance, appallingly destructive. And Albie would be getting the good news at about a metre's range.

But only if he tried to pick up the pistol.

The explosion was deafening. Rocking Mickey's world. Even with his fingers in his ears and his mouth open. He stood up and turned back to the door. The smoke that had blown through the doorway starting to clear. The office was a ruin. Glass blown out of the doors. The desk in two pieces, one on the far side of the room. The human wreckage that had been Keith smashed flat underneath it. A slowly spreading puddle of blood evidence that he was already dead. While Albie was barely recognisable as a human being. Having been catapulted into the wall behind him by the blast.

'He made his choice.'

Mickey nodded at James's blunt pronouncement. Still stunned by the havoc wrought by Albie's mistaken urge to take revenge.

'And now, I think, we should be on our way. Before your former colleagues respond to the 999 calls that I'd imagine the neighbours are making.'

36

The next evening. Angela was right on time. Mickey having decided to take the risk of Roz turning up to what had always been their favourite restaurant. Because when it came to fine dining in Monken Park, nothing came close to the Black Pearl. And if he was going to impress his guest, where better?

He stood up. And waited for her to make her way across the room. Turning heads and making tongues hang out. Metaphorically at least. At least one female diner giving her distracted partner the evils. The main feature of her outfit being a bright red midi dress. With a neckline that wasn't so much plunging as plummeting. Her secondary armament being a pair of what Mickey reckoned were Manolos. His expertise gained from long experience of Roz's weakness for shoes.

'Well, don't you scrub up nicely?' He kissed her on the cheek.

She looked at the table approvingly. 'Is that a martini?'

He grinned. 'Shaken, not sipped. Yet.'

'Very gentlemanly of you.' She shot a glance at the lurking waiter. 'I'll have the porn star version, please.'

Mickey waved the server away and helped her into her chair. Took his own seat, tempted to pinch himself.

'A porn star martini. You do realise this could be construed as harassment?'

'You just focus on showing a girl a good time, Mr Bale.'

'Mickey.'

'Mickey, is it? In that case I'm prepared to tolerate Angie. Not Ange. And definitely not Flange, in case you're ever tempted. Men have died for less.'

'I like Angela, as it happens.'

'And I like Mickey. I think it suits you. Michael for formal, Mickey for fun. As does the suit. It's nice to see you don't just have the same stuff you wear to work every day. What make is it?'

'It isn't actually a label. I had it made when I was coming through Hong Kong the last time.'

'Hong Kong? Where were you coming from.'

'Thailand. I lived out there for most of a year, after I left the Met.'

She raised an eyebrow at him. 'I hope you weren't there for either the children or the ladyboys?'

Mickey, tempted to reveal his year of martial arts training, decided to hold that back. Turned the gentle jibe with an even gentler riposte. Guided her to the menu's high points. Told her it was on him, and gave her free run of the wine list. Slitting his eyes at Paulo and telling him that the lady would be doing the choosing. And approved thoroughly when she chose a bottle just short of the pinnacle. Approving even more when he sipped from his glass.

'This is something else.'

'It should be. I chose it because it says grand cru on the label. A term that means of the highest quality.'

'Those shoes are pretty much grand cru.'

'These?' She stretched out her legs, provoking another round of poorly disguised staring from the neighbouring tables. 'They're what the footwear trade calls "fuck me" shoes. Which is why I wore them. So tell me, Mickey Bale, George Cross recipient and famous for your very own fifteen minutes, why it was you left the police when you obviously still love the Met? And no bullshit, I'll know if you're lying. I have my sources and they've come up with some pretty surprising theories about you.'

He looked at her for a moment. 'You were Met?'

'Seventeen years.'

'Then we might just have left for the same reasons.'

She shook her head. 'I left because I didn't like what my mates were turning into. You know, that whole four phases thing?'

He knew. A well-known meme, in the Met and probably beyond. Not applying to all officers, of course. Especially not the senior ranks, with everything to play for. But recognisable, nonetheless. Codified by some bright girl or boy. Then spread like wildfire. Propagating throughout the ranks via the twenty-first-century samizdat networks of WhatsApp and Instagram. Telegram for the more security-minded.

Its premise: the working life of the average copper dividing into four identifiable parts.

Phase one, years one to four. Baby coppers loving their new career. Away from home and exposed to real-life grit. Acclimatising to the impacts of man's mutual inhumanity. Amazed, appalled, astounded. A state of mind best summarised as 'fascination'. Everything new, and exciting. Lapping it up. Buying their own special kit. Clocking on early, going home late. And still believing in promotion on merit, bless them. The next two years: 'hostility'. Cynicism and disappointment. Most coppers getting second jobs to cope with marital penury or impending divorce. Hating and yet identifying with pushers, hookers and thieves. Doing the minimum, like disaffected footballers hiding from the ball.

The next nine years or so the sweet spot. Over the hump and into their prime. 'Superiority'. Coppers knowing how to play the game on the streets, in the nick and in court. Still fit enough to cut it, and handle themselves. The ones chosen to do the difficult and sensitive stuff. At the top of their game. Seeing the politics of promotion and managing to hide their disgust. Avoiding the unhinged and the brown-nosers. Sticking to their own circle of mates. Good coppers, but ultimately doomed to slide into the deep mediocrity of phase four.

The last fifteen years. Half a career. More like a half-world, for all but the precious few. 'Acceptance'. Most coppers now focused solely on retirement and pension. Seeking roles without danger. Custody or SOCO or Community. Avoiding the young and stupid. Anyone who might get them sued, disciplined, fired or killed. Longing for life on the beach. Constantly asking each other: 'What are you going to do when your thirty's up?' Old-fashioned street coppers like Deano, still up for a scrap at twenty-seven years in, as rare as rocking-horse shit.

'You saw your mates starting to circle the plughole?'

'More and more of them. Like their lives were going grey. And I thought, thirteen years of listening to that shit? Or worse, turn into that? No way. So that's why *I* left.'

He sensed her starting to reel him in.

'You think I was different?'

'I know you were. The rumours are pretty well circulated. Regarding a certain north London gang leader?'

Mickey shrugged. 'Rumours. The facts are that I got shot defending a principal, then hauled in on suspicion of being a vigilante. At which point the Met and I decided that we wanted a divorce. Mutual incompatibility, ill-health retirement, end of.'

The eyebrow raised again. 'Ill health? You look pretty healthy to me. In fact I'm counting on it. You can have two glasses of this lovely wine and no more.'

37

Eight thirty that evening. A time selected by the planners with great care. Close enough to rush hour for the tunnels to still be busy. But not peak period, with the associated risk of a jam. The ghosts under strict instructions to abort if needed. Not to risk their all-important clean getaway. Better to make another attempt than risk capture. But as they drove towards the tunnels traffic was flowing cleanly.

The team were in two cars. One for the device. A capacious late-model Range Rover with the rear seats folded. Tinted glass to prevent fellow drivers from getting an eyeful of the Sunburn in the vehicle's rear. The warhead held in place by the frame in which it had been shipped. The other vehicle intended purely for the escape phase. A ferociously powerful Audi estate.

Driving alongside each other, the two cars cruised the tunnel approach. Staying in line abreast as they descended into the left hand of the two tunnels. Traffic to front and rear all doing the same steady speed. A river of bored commuters heading for home north of the river. Some of them lucky. Those whose cars were in front of the ghosts' vehicles. Some others very much out of luck. So much so that even Ivan's customary black humour had deserted him. Knowing that he was about to commit mass murder.

They rode the slope down beneath the river at a steady forty miles an hour. Both drivers starting to slow, just a fraction. Allowing a gap to build to the cars in front. Ivan looking at Filip in the Audi from his passenger seat in the Range Rover. Asking the question. Got a nod back. Ready. The slope bottomed out at the bottom of the long climb back to the river's far bank. Ivan keyed his radio.

'Ready... three, two, one... *execute!*'

Filip braked hard. The Audi sliding backwards out of Ivan's field of view. Provoking a chorus of horns from the cars behind him. Forcing their drivers to stand on their brakes. Practically halting that lane's progress. Then accelerated hard before the drivers behind the Range Rover decided to go for the overtake into the momentary gap created. Using all five hundred horsepower to leave the car behind him for dead. Whipping past the bomb car in a blare of exhaust noise. Leaving a space into which Sasha promptly turned the Range Rover, parking it diagonally across the tunnel. Neatly closing off both lanes. Ivan and Sasha were out quickly, running for the Audi. Sasha pushing the remote central-locking button on the key fob. Ivan sending the radio signal that armed the Sunburn and started the timer. Rendering the car inaccessible. Its contents invisible behind the privacy film. And the warhead's detonator already counting down from 120 seconds.

Not that knowing what was about to happen to them would have changed anything for the people trapped behind it. The two men leapt into the Audi. Ivan snatching a glance at his watch. 110 seconds. Filip flooring the accelerator.

Bringing them up to the back of the traffic ahead in seconds. Merging into the pack. Knowing that the tunnel police would be scrambling to respond. Sixty seconds.

They emerged into the evening's pink dusk. Filip flicking a switch to light up the blue lights to front and rear. Hidden behind the grille and under the rear bumper. Pulling the Audi into the outside lane as commuters to front and rear obeyed conditioned reflexes and made way. Forty seconds. Gradually increasing speed as the traffic in front of them pulled over and let them through. Ivan looking at his watch. Counting down.

'Thirty... twenty... ten... just about... now.'

They were too distant to hear the explosion, deep under the Thames. But each of them knew what the effects of a Sunburn exploding in such a confined space would be. As predicted by the planning team. Two to three hundred killed in the initial blast. Incinerated in their cars by a compressed wave of fire. The lucky ones already dead from the overpressure wave that would have shattered windscreens into glittering shrapnel. Blasting the closest vehicles back up the tunnel in balls of fire. Perhaps a few would live, further up the slope. Those able and alert enough to abandon their cars and run for their lives. But the majority, it was predicted, would die from asphyxiation. As the tunnel filled with acrid smoke from burning petrol and rubber. At least another hundred dead. Probably a good deal more.

And beyond the immediate casualties, a targeting analyst's perverted wet dream. The bore would either rupture, allowing the river's water to flood it. Or more likely just fracture. The damage too severe for repair. With a sixty per cent chance that the other bore would be badly damaged too. Making it the perfect strike. Sowing fear on the country's roads.

Commuters realising that nowhere was safe. Reacting with panic to the slightest unusual road behaviour. And inflicting grievous economic damage as well. Costing the UK perhaps a hundred billion, all told.

Filip took them off the M25 at speed, heading south. Then killed the blue lights. Spearing down the A13 towards London for a single junction and then turning off into the side roads. They dumped the Audi at an industrial park. Whose CCTV had been taken down by hackers thirty minutes previously. Meaning that there'd be no record of the ghosts getting into the two cars parked for them only ten minutes before. Just as they were planting the bomb that had severed a major transport artery and killed hundreds. A seamless, perfect operation.

They headed back towards the M25, turning west to grind patiently through the inevitable delays. All four men quiet. Knowing the enormity of the blow they'd just struck. And that it changed everything.

38

James was first into the virtual briefing room. Still in Thames House, and summoned by a text on his work phone. Susan was next. Her face hard with the tension of unexpected action. 'I'm establishing an emergency conference. There seems to have been an attempt to bomb the Dartford Tunnel.'

Attempt? James stayed silent as she worked the technology. Andrew and Katy hurrying in as she opened the meeting on screen. Linking in the Essex police control room. A harassed-looking police officer appearing on the big screen. A conference call with MI5 probably the last thing he would have wanted.

'Hello, this is Susan Miles at Thames House. Please update us on the latest information with regard to your declared emergency at Dartford.'

To give the officer his due, he didn't miss a beat.

'Superintendent Clews, Kent Police. The latest we have is that a car was abandoned in the west bore just after twenty thirty hours. The occupants bailed out and departed the scene in a vehicle using illegal blue lights to clear a path. We think they made their escape via the A13 and presumably dumped the getaway car. Although there's no sign of it yet.

The abandoned car contains some sort of cylinder, from what our officers on the spot can see. We've warned them under no circumstances to touch the vehicle, and EOD are on the way from Ilford. ETA imminent.'

'Thank you, Superintendent. We'll keep this meeting open. Please advise as the situation changes.' She turned to the analyst team leaders.

'Andrew? No hints that this was coming?'

'Nothing. No chatter, no traffic increase. If this is a jihadi cell then it's one that we're not aware of.'

'Katy?'

Her head of intelligence shook her head.

'Nothing to indicate that there was anything like this brewing. Although it makes sense as target three in a campaign.'

'Expand on that?'

'Their opening shot was supposed to be a soft target. Borough Market again, at a guess. Hundreds of casualties, and in the name of Islam. The cathedral was a different direction, destruction of a religious antiquity, but still looking like it could have been jihadist. Now this...'

'Go on.'

'It's more calculated. Something the GRU couldn't resist dropping into the mix, perhaps. Something to make the boss smile. A soft target, but with high casualties and a huge price tag to put the damage right. A reminder that'll last until well past the next election. And it also feels like the penultimate target.'

Miles tilted her head slightly to one side.

'How so?'

'Two hits to set up the jihadi narrative. Both with named

Islamist groups taking responsibility. Then the third, this one, to misdirect. Make us reconsider everything we think we know. No jihadi involvement, no phone call to claim it.'

'So you think what, exactly?'

Spinner put her cards on the table. Going all in with impressive chutzpah. Either that, or with a big side order of monumental self-confidence.

'I think they've got one bomb left. And I think they'll use it for a proper spectacular. Something very public and very messy.'

The police superintendent interrupted her musing. 'We're getting a live feed from the scene, Thames House. Patching it through to you.'

The on-screen picture changing to a view of the tunnel. Shot from a hundred metres distant, a blue Range Rover front and centre. With a solid queue of empty cars behind it all the way back to the entrance. In the foreground a lone figure was walking down the incline towards the car. A long stick in one hand.

'Shouldn't he be wearing some sort of armour?'

James shook his head at Susan's question. 'It takes an age to put on, restricts mobility no end and it wouldn't be any use against a Sunburn warhead in any case. Discretion of the officer on the spot.'

The EOD officer got to within ten feet of the car, then raised the stick. The view switched to that of the camera mounted at its end. Looking down through the driver's window. A small but powerful spotlight shining into the car's rear. And putting the camera's image onto a head-up display on her smartvisor.

'Looks like ordnance to me.' Her voice clipped. Assured. And deceptively relaxed. 'I'm going for a better view.'

The view closing in. The car's back seats folded down to make space for a long tube. With carry handles clipped around the casing.

'Definitely ordnance. Potential Russian MO series warhead. Also some sort of timer detonator on one side.' The picture zoomed in tighter on the timer. Red rectangles seeming to float in the car's dark interior. 'Timer reading zero.'

Significant glances all round.

'Understood, Charlie Yankee. Return to forward mobile.'

All eyes turned to James. Who shrugged at his laptop's camera.

'I'm hardly an expert commentator. But if the timer reads zero then the detonator should have fired. I think we can count ourselves very lucky indeed. They'll send a robot in to put a hole in a side window and take a closer look. X-ray photography, most likely. But if that thing hasn't already gone bang it's most unlikely to. And I think you can trust EOD not to do anything to provoke it. That looks like a miss to me.'

39

Later, having insisted they go back to his, Angela renewed the questioning about Mickey's departure from the Met. Although not before mounting him in the hall for what she called the desperation fuck. And then again in the kitchen, pulling him in behind her while he was searching for wine glasses. Demonstrably grateful for his ability to renew matters so quickly. Mickey somewhat impressed himself. Opening her dress to allow him to cup her breasts while she pushed back at him. Finally making it to the bedroom, they had sweated and panted to a mutually satisfying conclusion. Talked idly for a while, both of them basking in the afterglow. Angela still probing, Mickey keeping it light. Then crashed.

Mickey woke in the dawn to find Angela drinking coffee beside him. Looking pointedly around the cramped bedroom before speaking.

'I'm pretty sure you didn't get thrown out for being on the take. I've seen bigger shoe boxes.'

He smiled back. 'The flat? It's a foot on the ground. I don't know what I'm doing with the rest of my life yet.'

She put the coffee down. Reached over and patted his

manhood. Cupped his balls and leaned in. Putting her breasts on his chest and her lips close to his.

'Never mind the rest of your life. You just need to be worrying about the rest of the morning.' Smiling at the inevitable reaction. 'Oh, *who's* a good boy? I could definitely get used to this.'

Mickey, lifting himself up to kiss her, caught a glimpse of his phone's screen. Switched to silent the previous evening and then completely ignored. Three missed calls. All from James. Shit.

He sent a holding text. Once he'd managed to persuade Angela that the missed calls weren't from another woman. 'Jimmy C' being a mate who was going through a messy divorce. The text promising to call back soon. Then returned his attention to the matter at hand. Not about to let something this good slip. Or at least not because he wasn't paying attention. After which they showered together. Breakfasted. And generally had a lovely time until Angela left to go and get changed.

'I'm on late turn again so it's time to go and get into the full bull dyke get-up. You on today?'

Mickey nodded. Chewing his toast and swallowing before replying. 'Yeah. Fourteen hundred departure from London City's private terminal. Monaco.'

'Shit, I'd completely forgotten it's a race weekend. Is he taking Mr Happy with him?'

Mickey guessing that she meant Kuzmich.

'Don't think so. Just me and Chef. Plus a plane load of sponsors and their wives.'

Angela drank the last of her coffee. Got up, pulled him close and whispered in his ear. 'I'm not worried about Chef, because you're not her type. Just stay away from the wives, all right? I've got a short-term let on you and I don't have any plans on a shared tenancy.'

Mickey promised solemnly not to do any Costnering. Kissed her out of the door and rang James. Who was evidently irritated.

'I'd prefer it if you were to return my calls a little faster if it's all the same to you, Michael. What on earth were you doing?'

'Entertaining, hosting, sleeping and then doing breakfast. Not that it's any of your business. What's the matter?'

'You've not seen the news then? A failed attempt by our favourite *domestic* terrorists to blow up the Dartford Tunnel.'

'Jesus.'

'Quite so. But the fuse failed to light the firework, meaning that a couple of hundred people are still alive. And the country doesn't have to find twenty billion down the back of the sofa to build another bridge.'

Mickey thought for a moment. 'And which leaves them in need of a spectacular.'

'Indeed. We probably have a few days now while they regroup and decide what to do next. You're still leaving the country this afternoon?'

'Yes. Back on Sunday night after the race.'

'Very well. I won't expect to hear from you until then. Keep your ears and eyes open for any clue as to where they might strike next.'

40

Mickey passed the flight to Cannes in pleasant discussion with Chef. The two of them having boarded Salagin's A318 last. Once the man himself and his guests were getting stuck into the Bollinger. Shown to seats in the euphemistically named 'entourage lounge'. Right at the front of the plane. Like the lobby to a flying living room. Two big boxy leather armchairs set at forty-five degrees to each other, occupying the cabin's left side. Faced by a series of wardrobes and equipment storage down the other side. Glossy wood abounding.

'Servant seats work for me. Means I don't have to put on a smile and make small talk with that lot.'

Chef's opinion, on being shown to the staff seats. Having both been there and done that. A veteran in the world of international high-end catering. Book read, T-shirt purchased. And therefore utterly unimpressed by any of it. She would, she told Mickey, infinitely rather be at home with her wife. If it wasn't for the five grand a day overseas duty allowance, of course.

'Flying out in this airborne wank tank is supposed to be some sort of privilege. Frankly I'd rather have flown business class with the rest of my brigade de cuisine. I think it's just to make sure that I don't go AWOL and start chasing skirt in Monaco. Given that us queers can't keep our hands off each other.'

Mickey pasted on his best politically correct non-committal face. Then laughed with her when she admitted she was only trying to wind him up.

'You're all right. We all thought you were going to be right up yourself, when we found out who you were. But even that hard-arse Flange seems to have warmed to you.'

Mickey nodded equably. Thinking 'if only you knew' and 'for Christ's sake don't let her catch you calling her that'.

The flight passed smoothly enough. Other than for a few minor pockets of clear-air turbulence over Lyon. Mickey, sitting with his back to the cockpit. The perfect view into the main cabin. And therefore able to watch events over his paperback.

He studied Salagin. Watching the Russian circulating through the opulence like a pro. Moving from sofa to sofa to press the flesh with everyone. Noting that he'd gone for the classic F1 insider's watch. Rolex Daytona of course. Platinum, with the giveaway light blue dial and chocolate ceramic bezel. A handshake here, a pat on the back there. Exchanging kisses in the Latin manner with his French and Italian guests. Chef smirked at Mickey knowingly. Reading the expression he thought he'd managed to conceal.

'Glad-handing, is he? Goes with the turf. Why else do you think he spent fifty million on this disgusting airborne display of conspicuous consumption?'

Mickey shrugged. Having decided he quite liked the Airbus's livery.

'Racing costs, big time. And these guys are his bankroll, I presume.'

She put down her tablet. Her face hardening.

'Yeah. His bankroll, our abusers. Keep an eye on them later, when they've all necked a bottle of fizz or two apiece. You might just save me from doing something Pavel Ivanovich will regret. Probably with my paring knife.'

'They get handsy?'

'They get cocksy too. I'll tell you a simple rule of thumb that'll stand you in good stead when it comes to picking protection jobs.' She leaned forward. 'Up to a hundred million pounds, or dollars, or euros, most wealthy people retain their essential humanity. North of that they turn into cunts, pure and simple. And there's nobody back there worth less than a billion. That, or they're married to the money and clinging on like a fucking limpet. Used to ignoring whatever he chooses to get up to because their prenups are too tight for them to bail in any sort of style. And they treat the rest of us like scum. We're just here to feed them, bring them drinks, and occasionally open our legs for them.'

She looked around, making sure that they weren't overheard. 'This fucking gig is a nightmare, year after year. I told my agency to up my overseas rate again this year, mostly to try and persuade him to get someone else to do it. But of course that's like telling him the two-hundred-quid-a-bottle wine is all gone, and he'll have to pay five hundred. Boo fucking hoo, get another one on ice.'

'It's that bad?'

'Like I said, they drink, they party, they do a few lines, and then they decide they want to fuck something. And being too lazy to go into town and too fucking mean to pay for it, they try to fuck the kitchen staff and the waitresses. I insisted on separate accommodation this year. We'll feed them, serve them, clear up and then get the hell out of Dodge. You and the male staff can have fun with them. And I wish you joy of that. Mind you, this lot aren't the worst. Just wait until the other team owners come on board.'

Mickey was first down the air stair. Putting on his sunglasses and wishing he weren't suited. Met at the bottom by a glossy-looking fixer. Early thirties, radiating the effortless confidence and health of the young. Long hair tied back in a tight plait. His one job to makes sure that the party transited from plane to yacht seamlessly. The reason for flying into Cannes and not Nice, of course. Even if an A318 was technically too much plane for the runway, at least in its laden tourist version. All for the priceless ability to park a helicopter within fifty metres of the stairs.

'Monsieur Salagin's party?'

Mickey turned and pointed at the red, white and blue Siberian tiger logo on the plane's tail. The A318 painted the same tasteful shade of almost French blue that had graced the team's cars since his takeover. Red already being taken, Mickey mused.

'Take a wild guess.'

Unperturbed, the Frenchman shrugged and spat a stream of incomprehensible instructions into his radio.

'The helicopter will touch down in ninety seconds.'

Salagin came down the steps with a smile.

'Bonjour Raymond.'

He watched approvingly as the baggage handlers sprang into action. A generous bonus generating unaccustomed speed and efficiency. Swiftly stacking his guests' luggage ready to load into the approaching helicopter. Mickey's concern as to the capability of one machine to carry fourteen passengers resolved by the machine itself. A Westland 101. The civilian version of the Merlin flown by the Royal Navy. More than big enough. Even with seats that weren't just glorified metal deck chairs.

Raymond waited until the machine had landed and throttled back to idle, rotors turning slowly. Then gestured to the cabin steward to send the passengers down the steps. Each couple pointing out their bags, neatly presented for identification. Which were then loaded into the big helicopter's cargo hold. Once all of the passengers were on board the machine lifted a few feet off the ground. Turned on the spot, the pilot undoubtedly showing off a little. Then slowly started to move forwards and climb away. Leaving Mickey and his principal standing by the now silent plane.

'Well done, Raymond. All we need now is…'

Salagin followed the Frenchman's pointing finger. A spot on the horizon rapidly resolving itself into a type of helicopter Mickey had never seen before. Painted the same blue as the A318. The two men watched as it flared to a perfect landing.

'For the guests, a rental will always suffice. They will have a slow and scenic flight, with a suitably cautious pilot. But this, Michael, is my own transport. And my own pilot.' The helicopter pivoted on the spot. Presenting them with the passenger door on the left side. 'Kamov 62. Made in Russia!'

Unable to stop himself grinning, Mickey buckled himself in. Finding Salagin's enthusiasm infectious, he strapped on the proffered helmet and plugged in the audio connection.

'All is good?'

He gave Salagin a thumbs up. Who in turn spat a choppy

stream of Russian at the pilot. Who grinned at them over his shoulder. And launched into the gunship pilot's tour of the Côte d'Azur.

'Alexei was a navy pilot for fifteen years. Makes him the perfect man to fly my brand-new chopper! What do you make of the 62, Alexei?'

The pilot laughed. 'I spent fifteen years flying the 27 in the Northern Banner Fleet. A design that was already at sea by the time I was born. Built to carry torpedoes and a dipping sonar. Fifteen years hunting for the American attack submarines in a flying tow truck. This is like riding a racehorse after taking the saddle off a cow!'

He piloted them expertly out to sea at low level. Skimming the waves at no more than fifty feet and turning east. Salagin pointing out Nice and Cap Ferrat on the left side. Threading through the yachts moored opposite the hilltop village of Èze. Its rooftops hundreds of feet above them. Mickey busy working out whether to vomit or not. And managing to not do so by the skin of his teeth.

He sighed with relief as Alexei dropped the Kamov onto a smartly painted flight deck. Took off his helmet as directed. Handing it to the co-pilot who was holding the passenger door open for him. Finding himself on the roof of what had to be the biggest yacht he'd ever seen.

The boat named *Saris*, Mickey knew from his idle research; 534 feet long and capable of 30 knots, flat out. Salagin's latest ego extension, replacing an inferior version. Which had apparently lacked guest facilities. And, Mickey suspected, a proper suite of defensive aids. Laser missile jammers. Chaff dispensers. And quite possibly a close-in defence weapon. A pop-up gatling gun, it was rumoured. Radar guided for those last-resort moments. When spraying high-velocity exotic metal slugs by the thousand might be the only answer to whatever was incoming.

Saris being painted, of course, the obligatory almost-French blue. Salagin took a pair of sunglasses from a tray held by a uniformed crewman. Picking from a choice of blue, green or brown lenses. The Russian selecting the blue, predictably enough. Then handed Mickey a pair, choosing the same colour for him. Top-of-the-line Persol, of course.

'I like everything colour co-ordinated. Which is why I suggest you go to your cabin and see what's waiting there for you. And welcome to *Saris*, Michael. I think you're going to enjoy this.'

41

'It's something of a tradition.'

Two hours later. The afternoon sun dipping toward the horizon. Chef's team setting up the buffet under her eagle eye. Two male servers waiting behind a magnificently stocked bar. And Salagin, master of all he surveyed, sitting in an easy chair. Legs crossed, leaning back into the cushions. Sipping a Pimm's and totally at ease.

Mickey had changed for the afternoon on arrival, as suggested. Agreeing that his usual suit wasn't either going to be comfortable or portray the right image. Led to his cabin by a crew member, he'd found three sets of clothing waiting. All subtly different. Light blues and off-whites to match the yacht's colour scheme. Of course. Along with perfectly fitting deck shoes, slipper socks and a crisp new Panama hat. With a perfectly fitting blue all-wool blazer for later in the evening.

Freshly attired, he'd made his way up to the rear deck. And asked if it was OK to use the big binoculars bolted to the handrail over the stern.

'Of course. They're image-stabilised.'

And so he'd spent a careful ten minutes scanning the shoreline. Looking for something very specific. And then, even though he'd found nothing, made a phone call to James. Then nodded to himself and walked back to join Salagin.

'You were looking for…?'

'Anything that mounts an expensive piece of glass. When we flew in I got a good look at the dockyard. Lots of places for someone to hide.'

Salagin was admirably unruffled. 'Expensive glass? Are you thinking cameras or rifles?'

Mickey shrugged. 'Either. You're not exactly the lowest-profile target either way.'

The Russian nodded, untroubled. 'Que sera, sera, Mr Bale. The clothes suit you, by the way.'

'Thank you. For the clothes, I mean.'

The Russian waved a hand. 'I like my people to fit in. It makes them feel at ease.'

'And more likely to do their jobs right?'

Salagin shrugged. 'You'll see my point when the other owners arrive. Speaking of which…'

He gestured to the harbour. A speedboat was carving through the water toward them. And gestured to the platinum Daytona. Two minutes to the hour.

'That'll be Sinclair. He has a pathological hatred of lateness.'

Mickey knew who Sinclair was. Along with every other F1 fan. Ross Sinclair. Billionaire owner of the recently renamed Racing Supremacy. The RS motif no surprise to anyone familiar with the size of the man's ego. The rest of the pit lane seeing the name as a massive hostage to fortune. Only Ross seeing it for what it really was. An open declaration of war. Having bought one of the most promising teams on the grid. And then proceeded to throw money at it. Buying the best of everything that was for sale. The hottest design team.

Facilities to match their ambition. And experienced spanner wielders from all the way up the pit lane. Predicted to rival Team Rossiya for best of the rest within a season or two. Racing Supremacy also being home to great drivers at both ends of the spectrum. One of the better newcomers, who also happened to be Sinclair's son. The reason Ross had become interested in the sport in the first place. And in the other seat, Nathan. One of the greatest drivers in the sport's history. Inexplicably let go by the championship leaders at the end of the previous season. And lured to Supremacy by a staggering amount of money. Plus major contributions to his pet charities. The man who had stepped out of his jet in Wiltshire a few days before. To talk about moving to Rossiya.

Nathan clearly having become somewhat disillusioned by Sinclair's monomania. And totally pissed off by unexpected team orders. An edict that his teammate was the priority. And that Nathan was expected to surrender track position on command. Whatever necessary, to get the kid up the pecking order. And then to play wingman for Ross Junior. Keep the bandits off his tail.

Not the sort of thing the three-time world champion thought he'd signed up for. Nothing contractual, of course. Because if there had been he'd never have signed. Just relentless pressure from Sinclair, through his hapless team principal. Want the aero upgrades? Want the engine parts? Want reliability? Then do the sensible thing. Not the cleverest way to manage a world champion. Hence Nathan's very well publicised dalliance with Rossiya.

The boat slowed to come alongside. Mickey leaning over the rail to take stock of the occupants. Sinclair up alongside the driver. And looking like he wanted to be the one at the controls. His wife sitting in the back. Looking, it had to be admitted, stunning. What had to be a bodyguard alongside her. Horribly out of place in what looked like a standard-issue US Secret Service suit.

They boarded. Salagin greeting Sinclair with the savoir faire of an experienced philanderer caught in bed with an associate's wife. Sinclair having clearly decided to display the same nonchalance. The two men exchanged pleasantries. Sinclair striving to avoid looking like he gave a shit about the Nathan thing. Mickey exchanged professional greetings with the Secret Serviceman. Nods at twenty paces. Amused to see the guy had the clichéd transparent curly wire connected to an earpiece. And wondered what he could do with it. Call in an airstrike? Order in pizza?

Several more of the most powerful men in the sport arrived over the next half hour. And Salagin's own world champion Kurt made an appearance from his cabin. Evidently still well regarded enough to merit a place on the boat. Mickey marvelling privately at meeting men he was accustomed to viewing through Netflix's all-seeing eye. And at seeing them chat as if they hadn't a care in the world. Or weren't bitter rivals. Although any underlying bitterness was probably amateur status. At least when compared to the grief brewing between Sinclair and his own principal.

Seeing the confrontation building, he relocated, little by little, closer to the Russian. Keeping a careful, if discreet watch on his American counterpart. Sinclair sinking glass after glass of Dom Pérignon. At length putting his drink down and squaring his shoulders. His previous attempt at sangfroid evidently shredded by the wine's intoxicants. Ignoring his wife's pleas to quiet down, and calling out to Salagin.

'Can a guy get a beer round here? Hey, Pavel, you got any good American beer, son?'

Salagin smiled his inscrutable smile and gestured to the barman. Who brought a bucket of beers out and placed it on a table beside the American. Sinclair took one by the neck and looked at it in disgust. Then drained it in one long extravagant swallow. Grimacing in disgust.

'Fuckin' Heineken? What sort of horse piss is that to offer a guest?'

Mickey looked at his bodyguard. Who was talking fast to his cuff. Either calling up their boat or just adding more anchovies to his order. But not doing what he needed to be doing. What his principal needed most from him. Which was recognising that the boss was at the point of losing self-restraint. And finding a tactful way to usher him away. Before something newsworthy, and possibly litigable, happened. Perhaps telling him that there was a bucket full of Bud on the ship's stern. Any blatant lie that would work, basically. Anything to avert what Mickey suspected he was about to have to deal with.

Time slowed a little. Mickey's senses flicking into combat mode. As Sinclair put his hand on the neck of another bottle. But not the grip he'd have taken to drink from it. Rather that best suited to throwing. Mickey fast-forwarded the next ten seconds in his imagination. Bottle thrown: multiple outcomes. Broken glass: not good, but not catastrophic. Salagin hit by the bottle: nothing good about that at all. The consequences possible serious injury and unavoidable dismissal of his protection officer. Or the worst potential outcome of all: another guest hit by the bottle. Given there were half a dozen team owners' wives in the field of fire. He moved. Fast.

Stepped in. Taking hold of the bottle. Just as Sinclair drew it back over his shoulder to throw. Pinned it there. Upside down, given the American's hold on the neck. Resulting in the contents cascading down the team owner's highly expensive jacket. Took the empty bottle from him, and—

Secret Serviceman came in from behind his principal. Having used Sinclair's considerable bulk to disguise his approach. Although Mickey, combat-conditioned, had known he was coming. Hearing the tapping of his Brooks Brothers heels on the wooden deck. And had factored it into his impromptu battle plan.

Only needing to know which side of the big American the bodyguard was going to pass. Even had the time to toss the empty bottle lightly to the nearest server. Who, to give him credit for a cool head, caught it cleanly.

The bodyguard choose to come around his boss's back. Wise, given Sinclair looked like he was getting ready to punch someone. Mickey being the main candidate. But that wasn't going to happen for two, maybe three seconds, given his state of intoxication.

He improved the odds of that punch taking too long to deliver with a gentle push. Both putting the big man off balance and clearing the space he was going to need very shortly. Sinclair tottered away. His sense of balance perhaps a bottle and a half of fine champagne on the wrong side of fully functional. And the bodyguard attacked.

Or rather, tried to attack. Mickey stepping back and swaying back at the waist to allow the punch to pass him harmlessly by. But knowing the other man wasn't going to let that put him off. Considered breaking the extended arm for a split second. Deciding that, as making a scene goes, perhaps a little on the side of over the top. The screaming, for one thing. The ambulance helicopter for another. And decided to go for something just a little more subtle.

Having stepped smartly back and to his left, he had adopting a balanced stance. Good for fast movement, in case the other guy was a fighter too. Deliberately putting himself to the other guy's right. His right arm drawn back across his chest. A stance with options, arm breaking the most extreme.

'Nine times out of ten you're going to want the other guy asleep at your feet is all.' Chuck, one Thai morning so humid that it was all Mickey could do to breathe. *'He's... what did you call it... ah, yeah, fronted up. Leaving you with no option but to fight. So, what you gonna do? Sure, you can bust his jaw. Snap his wrist. Half a dozen options. But you do that stuff, you're likely to end up in front of a judge for excessive force. Save that for when there's too many of them for you to go easy. So here's what you do...'*

He chopped at the side of other guy's neck. Fingers bent to tense the side of his hand into stiffness. Hitting the vagal nerve dead centre. Stepping back as Secret Serviceman crumpled at his feet.

'What did you fucking...?'

'Ross!'

Sinclair's wife. Her patience exhausted. And exactly the sort of Iron Butterfly required to deal non-violently with her husband.

'*You* did this! That poor man was only stopping you doing something really stupid. Now get your dumb *fuckin'* ass down on that boat before I take it and make you swim back!'

Mickey, tempted to applaud, refrained. Not wanting either to attract her ire or draw any more attention to himself. Called a couple of crewmen over to carry Secret Serviceman down to the stern. Watching impassively as Sinclair beat an undignified retreat. Face beetroot red. His parting shot the harbinger of more grief to come.

'I ain't done with you, you Russian sonofabitch!'

Later, with the guests departed and the buffet cleared away. Salagin patted Mickey on the back and declared himself happy with the outcome.

'Even if I managed to make him even more pissed off than he already was?'

The Russian shrugged. 'He was already beside himself with the need to save face. Being rejected by a driver like Nathan isn't an easy thing to weather. And even if it was all his agent's doing, Sinclair will paint himself as the aggrieved party. This one, as they say, will run and run.'

Mickey looked at the harbour's lights. Thinking for a moment. 'What if I had a way to smooth all this away? Give you some leverage on him?'

Salagin raised an eyebrow. 'And do you?'

'I might have. It'll cost you, if it's available.' He raised his hands. 'And none of it'll go into my pocket.'

The Russian smiled that ineffable smile. 'You have me interested. Define "cost".'

Mickey shrugged. 'I don't know the market. Give me a minute and I'll see what the SP is.'

'The... SP?'

'Sorry. Slang. Starting price. It's a horse racing term.'

He walked away. And dialled a number that wasn't so far down his recent calls list. It rang twice.

'Fuck off, Mickey. I'm just talking to someone. I'll call you back.'

A decent start in itself. She was still taking his calls. Despite probably being pissed off at him for lying about his new role. 'Whatever they're offering, forget it. Think about paying your mortgage off instead, eh?'

'*What?* You can't be seri—'

'Yes, I can. Come out from your hidey hole on top of that container and walk down to the jetty. He'll probably want to meet you. And bring your gear with you. He'll probably let you do the full snapper's tour round his new blue toy.'

She was silent for a moment. 'How did you know I was here?'

'People I work for these days are pretty good at locating mobile phones. Now do you want the money or not?'

42

'Delta Three One, Roustabout is out of UGS and mobile. Black taxi, index Alpha Yankee Six Seven Charlie Echo Foxtrot. Proceeding west Upper Grosvenor Street. Yankee One One can?'

Kev translated automatically. Having learned to run Service radio speak in real time. The static observer, Delta Three One, was watching Salagin's Upper Grosvenor Street residence from his place on the pavement. Huddled in a sleeping bag, reeking of piss. Taking authenticity to a new level. And had called out that Salagin's associate Yuri Kuzmich, tagged as 'Roustabout', had left the building. Climbed into a black cab and was on the move. Heading west toward Park Lane. And was asking if Kev, waiting on his bike in Park Street, was able to take up the follow. Helpfully providing the cab's vehicle registration number.

'Yankee One One can.' Kev flipped down his Shoei's face visor. Thumbed the bike's starter button and pulled out behind the taxi. 'Yankee One One in the follow.'

Kev was having the time of his life. Astride the perfect town bike. A BMW R stripped of its fairing. The flat twin layout making for a low centre of gravity. The Bimmer chuckable in a way his previous Job bikes just never were. But then his previous role as a Special Escort Group Easy Rider had been imperious. Rolling in front of the VIP cars, regulating convoy speed while his colleagues on the Working Bikes did the dashing about. Alternating vanguard and rear-guard.

Kev had loved the Working Bike role well enough, of course. What wasn't to love? Facing down oncoming traffic with a whistle and a hard stare from behind a smart pair of sunglasses. One hand holding the bike, the other on the pistol at his waist. Authority personified. But Easy Rider? That was true power. Shepherding the great and the not so good across the capital had been what he was born for.

Truth be told, the shrapnel that had nearly killed him had done him a huge favour. Not just the medal, although that was a nice end to his career. But getting him noticed by MI5. Picked up, dusted off and put back to work. And not so much work as playtime.

Running urgent dispatches, documents not to be trusted to the network. Discreet transport of personnel in one of the Service's black cabs. And, best of all, the occasional follow job. When the field teams were overstretched.

And right now the field teams were at the ends of their tethers. Tasked to watch a hideously complex and varied set of would-be terrorist cells. Anywhere there were Muslim immigrants in any number, there were suspect groups. Many of them fantasists and blowhards. And very few of them posing any real risk. But all of them having to be tracked 24/7. Given the PM's promise to the nation that there could be no repeat of the last two terrorist outrages.

And it was worse than that, as Kev unofficially knew. Having heard on the grapevine that a device had somehow failed to detonate in the Dartford Tunnel. With, so the gossip went, at least one more Sunburn out there. Which was why he was on duty well after the usual close of business. Riding a very careful follow on a target of interest.

The cab merged with the southbound Park Lane traffic. Took Hyde Park Corner and proceeded west with Kev in loose trail. Keeping a couple of cars between himself and the cab. Trusting the black helmet and leathers to keep him more or less invisible, even if Roustabout was alert to a follow. The bike's upright position allowing him to look over the top of the car in front.

The invitation for him to come and play had been pretty last-minute. The briefing swift and business-like. Roustabout being suspected of being a Russian illegal. The go-between for the strike team currently bombing the shit out of the UK mainland. Their last attempt having failed only because of a component failure in the warhead's detonator. Good fortune that was unlikely to be repeated.

There was a pattern, it seemed, of Roustabout going missing for an evening with each fresh attack. First noted before the cathedral bombing, by an asset in place inside UGS. And confirmed, belatedly, with the failed tunnel attack. Which meant that stretched resources or not, he had to be tracked to his destination for the evening.

Which would immediately be staked out by the team's other resources. A pair of follow cars with three operators apiece and another bike, Sierra Two Five. All cruising behind Kev in the early evening traffic at a safe distance. More assets available if Roustabout, as expected, ducked into wherever he was going and promptly headed out of the back door.

And for this very special evening, the big guns were in

attendance. Special Forces on standby, further back in Op Revoke's order of battle. Two supercharged Range Rovers, each four up and weapons-heavy. Ready to follow Roustabout wherever he went. With gunship support on the menu, if needed.

Making Kev the point of a very expensive and highly capable spear. The one man keeping all that expertise and firepower in touch with the target. He looked in his mirror, seeing the bike closing up on his rear. Easy to spot, white helmet standing out in his mirror. Surprised, but cool with Sierra Two Five moving into position ready to relieve.

He put his thumb on the transmit button, ready to make the handover. *Sierra Two Five can?* Then allow the other rider to ease past him, take the next left, two rights and a left to fall in behind. Realising just a second too late that Sierra Two Five was wearing a black helmet, like his. And not a white one. The other rider clearly in a hurry, swerving left to undertake Kev's cruising BMW. Something Kev had always hated with a passion. Moved his thumb to the horn. Ready to give the prick an earful as he whipped past.

And suddenly found himself sprawled over the bike. Having landed hard on the wrong side of the road. Still sliding, foot peg, handlebars and tank grinding as they dragged against the tarmac. Craned his head to look up. Into a red wall.

43

Mickey got back to the UK late Sunday night. It being an article of faith that Salagin never stayed over after the race. Celebration or commiseration both matters for the team and not the owner. Unless they actually won, that was.

Salagin himself being pretty pleased, all things considered. Rossiya having come home in decent points-scoring positions. George, gaining in confidence with every race. And stealing fourth from an ailing McLaren on the last lap. Kurt, of all people, cutting a delighted figure. Striding up the pit lane to congratulate his teammate. Having been his rear gunner for half the race. And penned a gaggle of frustrated mid-table cars up behind him in the pencil-thin street circuit for five critical laps. Stopping them from getting after George. Who had taken the chance with both hands. Pushing the gap between them hard enough to achieve a flawless over-cut. Pitting for his second set of tyres with so great a lead on them as to be uncatchable.

Kurt's new-found serenity in the face of the new kid in town not unnoticed by the cognoscenti. Who wondered breathlessly where that had come from. Mickey reckoning it was the result of a late-night chat. While Tamara had been taking the *Saris* tour, Kurt had wandered up, beer in hand and wearing a friendly expression.

'I watched what you did with Ross. That was some impressive shit, man.'

Mickey had nodded his thanks. Smiling at the German's next question. Realising this wasn't about him.

'But how do you stay that sharp at your age? I mean, no diss intended, but you must be...'

'Well into my forties. Yeah. I have a secret recipe.'

Kurt's eyebrows rising. 'You care to share?'

Mickey's secret recipe being part Prot wisdom, part Chuck-inspired serenity. Chuck being a bit of a combat psychologist on the side. Peddling meditation and inner calm alongside one-punch knockouts

'It's easy. Stop giving a shit.'

'What?' Kurt clearly still very tightly wound, behind the relaxed demeanour.

'It's wrapped up in a load of mysticism. And it takes some practice. But at the heart of it, it's just about letting go. Knowing that nothing really matters.'

'Right...' The German getting ready to disengage.

'And one more thing. Knowing that *you* don't matter.' He fixed Kurt with a stare. Not hard, or challenging. Just truth. 'Nothing about *you* matters, Kurt. Or me. We're both what life made us. That and the people who helped us get to where we are. After all, I remember your first victory. Belgium, right?'

The German smiled. Liking that memory. Winning at Spa. Shepherded to the victory, as Mickey recalled, by his older teammate. Who, coming out of the usual hectic Spa hurly-burly at the start behind Kurt, had opted to run interference for his teammate. Rather than trying to muscle past him. The result the unleashing of a phenomenal talent.

'Could you have won that race without Enrique keeping the Ferraris off your tail?'

Kurt nodded. Then gave Mickey the stare back. As the truth of it struck him. 'Are you giving me career counselling, bodyguard?'

Mickey grinned. Knowing he was two for two already. And giving no shits. 'Perhaps I am. You can either listen and think, or you can take offence and keep on down your current path. Pushing too hard because there's a kid up your arse. Punting him off, putting yourself in the gravel. You know where that ends up. Or you could give Enrique a call and tell him thanks again. Perhaps get the same call in fifteen years from George? If your ego will let you.'

Kurt nodding stiffly and bidding him goodnight. And walking away, thinking. The result? Mickey Bale, unbelievably, three for three.

Salagin's other source of satisfaction had been the absolute silence from Racing Supremacy. An uncharacteristic 'no comment' from Sinclair on the rumours that had led to the suspension of betting on Nathan's departure. A silence engendered by Salagin having emailed over a set of pap shots. Long-lens work of the highest quality.

Which clearly, and frame by frame, showed the American being prevented from throwing the beer bottle. Accompanied by a one-line message: 'We can have peace, or these can go public. It is entirely your choice.' The inference being that Sinclair needed to pick one of the two outcomes. And, it seemed, picking peace.

The two team owners had seen each other on several

occasions over the weekend, obviously. And Mickey had readied himself for round two. Tensing the first time he saw that distinctive suit. Only for Secret Serviceman to stroll over. Hands raised, open palms forward. And, somewhat to Mickey's surprise, nodding respectfully.

'I'm Jack, and I know you're Michael. Good to meet you without being put to sleep. And no complaints here, I was totally off my game. I mean, props to you, buddy – that was pro-level work. But if I hadn't been so wound up by the idiot's drinking I might have been a little savvier, you get me?'

And Mickey had nodded graciously. Having no doubt that he'd rather have Jack as a friend than an enemy.

'Oh, and nice watch, man. Been looking at one of those myself.'

Jack eyeballing Mickey's new Rolex Daytona. The black face, of course. Subtle and understated. Bought for him the day after by Tamara. An unexpected share of the proceeds from her unexpected sale of the Sinclair photos.

'You just scored me enough money to buy a house with, Mickey Bale. And I believe in sharing the love.'

Once aboard *Saris*, initially suspicious, she had swiftly been charmed by Salagin. And had happily abandoned a £50,000 UK media offer for the photos. Chicken feed, in the face of a mind-boggling half million from Salagin. That, and a behind-the-scenes photo tour of the yacht probably worth another hundred k. Salagin inviting her to stay on for a week once everyone had left Monaco. An offer she had accepted with alacrity.

'Apparently it's going to be cruising to his next planned holiday location. Which is the British Virgin Islands! Too bloody right I'm staying aboard!'

Mickey had returned to the UK aboard the flying living room a contented man. His job performed with professionalism. His principal very much indebted to him. And his advice heeded by a man he'd previously seen as a demigod. And walked straight into the teeth of a very chilly wind indeed.

44

Mickey walked down the air stairs at City. Last off the plane, other than Salagin, who was thanking the pilot. Finding Kuzmich waiting at the bottom.

'Michael Bale. You are quite the hero of the hour, I hear.'

Mickey shrugged. 'Sometimes you get the bear.'

Expecting Kuzmich to smile. But disappointed. The Russian's delivery of the traditional punchline deadpan. Perhaps even a little hostile?

'And sometimes, Michael Bale, the bear gets you.'

Mickey sensed disapproval. Perhaps something stronger. And was tempted to do the old 'you're welcome' gripe. But restrained himself as Salagin came down the steps. Kuzmich looked up at his boss. A non-verbal message in his neutral stare. And spoke to Mickey without looking at him.

'You are dismissed, Michael Bale. On duty tomorrow morning, yes?'

'Eight o'clock sharp, Mr Kuzmich.'

The Russian shook his head. 'There will be no need for an early start. You can come on duty at ten thirty and it will be plenty of time.'

'Ten thirty hours it is. Thank you. Goodnight, Pavel Ivanovich.'

'Goodnight, Michael. And thank you again.'

He could tell Salagin sensed it too. Saw the hint of uncertainty in the other man's eyes. Shook his outstretched hand and headed for the private jet terminal. Eager to get through immigration and give Angela a call. Surprised to find a text from her waiting on his phone.

'Cross the roundabout and go down Thames Road. Blue Golf. Make sure you're not followed.'

45

Salagin watched Mickey walk towards the terminal. His expression of bonhomie sliding into a neutral mask.

'I presume from your treatment of Mr Bale that you have bad news?'

Kuzmich nodded. 'Bad enough that I could not trust it to any form of communication open to surveillance. But more importantly, the Watchmaker has issued the order we were expecting.'

He watched Salagin's face intently. Relaxing a little as his friend nodded without any change in his expression.

'It was only a matter of time. The arrangements are all in place?'

'Yes. You sign the papers at eleven thirty, and the jet will be waiting.'

'And the last *Solntsepyok*?'

'Is scheduled for delivery at the same time. Everything is in place.'

The oligarch nodded. 'I thought I would miss all this. But of late, I find the mask beginning to slip, just a little. It will be a pleasure, in a way, to live a simpler life. Not that I have any choice in the matter.'

46

Mickey crossed the roundabout outside the terminal.
Still surfing a wave of goodwill to all men. Glancing
behind him before walking down Thames Road. Seeing the
blue car. Delighted to be met, wondering what was behind it.
Tapped on the window. Smiled as she wound it down.

'Hello hello hello! What's all this then?'

His smile faded when he saw the look on Angela's face.
And her finger at her lips. She held up her phone. A line of
text in an unsent message for him to read.

'GET IN. SAY NOTHING.'

He put his case in the boot and got into the car. Watching
her as she drove them back into London. And realising that
whatever it was she wasn't discussing with him wasn't good.

47

'And Bale?'

Kuzmich turned to look at his friend as they walked towards the terminal.

'The sensible thing is to take him out of the picture. I have issued instructions to a third party.'

'I don't want him killed.'

'You're becoming a sentimentalist, Pavel?'

Salagin shrugged. 'In another life you and I could have fought alongside him. He has a strong sense of justice. And he has done the right thing for me. So no fatality. Just put him out of the game for a while.'

'As you wish.' Kuzmich took out his phone. 'Why don't you complete the formalities while I emphasise that to our chosen contractors?'

He waited until his friend was inside the terminal before calling his contact.

'Hello... yes, me again. I have fresh instructions for you with regard to the target. You are instructed to make sure that the action is non-fatal. I repeat, non-fatal. Is that understood? Yes? Good. Now listen very carefully. You are to pass that instruction to your operators. And then you are to tell them that I expect that there might well still be an accident. A fatal accident. Perhaps the target will fight back too effectively for a non-lethal outcome. For which there will be no punishment. Do you understand what I am telling you?'

He listened to the reply.

'Yes. I think we understand each other perfectly. And I will pay double, just to ensure that the job is done perfectly. Make it happen.'

48

An hour later. Mickey and Angela in a restaurant of Angela's choosing. Hidden down a back street in the city. The Golf parked handy for an exit through the rear. And her phone linked to the car's dashcam. Allowing her to keep an eye out for anything indicating trouble. All of which was freaking the shit out of Mickey. His Monaco-inspired feeling of goodwill totally evaporated. And replaced by a general feeling that all was far from well. Each of them with a drink in front of them. Neither of them drinking.

'So...'

She looked at him across the table. Reached out and took his hand.

'I'm sorry, Mickey, I've got some bad news for you.'

'Is it about us?' Mickey knowing it was all a bit too cloak and dagger for a break-up.

'No. It's about your friend Kev.'

'Kev? How could you know...' Cogs turned in Mickey's head. Powering the inevitable light-bulb moment.

'Wait. *You're* Box?'

'Yeah.' She stared at him. Full-on apology in her eyes. 'I'm sorry. If I'd known you were too I'd have... I don't know... not got into this, perhaps?'

It being obvious to them both that entanglement of two operators wasn't such a great idea. Mickey getting over the apparent breach of protocol effortlessly. Since he didn't actually care about the rules.

'Bit late to worry about that. Not that I am.' He squeezed her hand reassuringly. 'I mean obviously it's down to you as well, but for me we're in "oh dear what a pity never mind" territory. So what about Kev?'

'He's dead.'

Mickey struggled with that for a moment. Badly. And was nowhere near being past disbelief when he asked the next question.

'How?'

'He was hit by a bus. While he was following that bastard Kuzmich on his bike. Kuzmich was in a cab and keeping an eye out, so the closest follow car was two hundred metres back. Which meant they didn't see it happen. But the bus cameras caught it all.'

'And?' Mickey's anger starting to tighten its scaly grip.

'Someone rode up behind him on another bike and just pushed him off. He went under the bus coming the other way. Never had a chance. DOA.'

'So it was quick?'

'We think so.'

Mickey was silent for a moment. Making a mental note for himself. Marking Yuri Kuzmich down as dead. The bastard just not knowing it yet.

'We?'

'I work for Tom Stoddard in Field Operations. Your

handler didn't know about my presence on site until yesterday. Chinese walls. When Susan Miles decided to pull us both out.'

Mickey thought about that for a moment as well. One part relieved. One part aggrieved. And aggrieved put relieved straight back in its box. 'That's not her choice.'

Angela sitting back. 'They think that there's another strike inbound. And that Kuzmich shook off Kev because he was on his way to brief the Spetsnaz team. On top of that another extremist cell has been woken up. Their intention seems to be to travel to London tomorrow. And it's assumed that the Russians will hand the warhead over for them to martyr themselves with somewhere nice and public. So the plan is to follow them, with Special Forces ready to strike. Hit them the moment that the Russians show up. And then to hit UGS and the GazNeft offices at the same time. Maximum attack. So they want us well out of the way.'

Mickey shook his head. 'No-one's settling the score with Kuzmich except me.'

'Mickey...'

'What are they going to do? Nick me?'

She shrugged. 'It's possible. One word to the Met and you wouldn't see the light of day until it was all over.'

Mickey thought about that. 'We need to talk to James.'

He dropped a note on the table to cover the drinks. Plus a generous tip. Apologising to their server – something had just come up. They exited the restaurant into the evening chill. Turning left to go around the block to Angela's car. And Mickey's copper-trained street sense pinged hard. His endocrine system making the unilateral decision to deliver a prompt dump of adrenaline.

Two men emerging from the shadows in front of them,

twenty metres out. He took a quick look back. Another two waiting on the far side of the restaurant, thirty metres back. Making any thought of flight a waste of effort.

Angela, still moderately relaxed, realised he'd stopped walking. 'What's the—'

'Stay here. Call 999. We're being mugged.'

Already pretty sure that there was more to it than that. He disengaged her arm, gently pushing her away. Strode forward to meet the two in front of them. Knowing he'd have to make it fast, to avoid having to fight off four men. They were blocking the pavement, grinning easily. Young, mid to late twenties. Fit, toned even. And confident. Strength in numbers. Perhaps some fight training too. One a good three inches taller than the other. Slightly closer to Mickey as he walked towards them.

'Watch and wallet, mate. You know how it works.'

A flicker of light at his side. Street light reflection off polished metal. But not brandished as leverage. No overt threat. Intended to be used without warning. Mickey's mind slipped from calculation into automatic. Muscle memory from a succession of roasting-hot days. When each sparring session had left him gulping litres of water that never reached his bladder. Chuck having decided to teach him how to deal with knives.

'*You're gonna face this, sooner or later, right? You Brits don't do guns all that much, but you sure do blades from what I read. And let's not forget that thirty per cent of all stabbing victims die.*'

Mickey pretty sure that the intended ratio for this attack was more like a hundred per cent.

'*Most knife attacks use a hammer strike, straight down or*

diagonally. And it only takes an inch or two of penetration to kill. So here's how you counter that…'

Tall Guy moving in. The knife still held facing backwards. Supposed to be invisible. Mickey knowing he was lucky to have caught the glint. Unlocking any restraint he might have exercised.

'You sure about this, boys? Last chance…'

Mickey stepping to his left. Closer to the shopfront, a small tactical disadvantage, but putting Tall Guy between himself and the other one. Tall Guy deciding to go for the kill with the impetuosity of youth. The knife coming up in a stab for Mickey's gut.

Mickey blocked left-handed. Pushing the knife hand wide. The adrenaline making his movements a well-practised blur. Reaching in with his right over the top. Gripping the knife hand and bending it back, locking the arm up. Tall Guy suddenly too focused on fighting for control of the weapon to do anything useful with his other hand. Mickey lifted his arm violently, putting his opponent off balance. And then, knowing he only had a second or two, used the momentum of a pivot to twist the wrist savagely. Breaking it instantly. Stepped back to reset. Ignoring Tall Guy, who was staggering backwards staring at his ruined wrist.

Other Guy, to give him credit, didn't hesitate. Came in fast, his blade swinging at neck height. Going for a killing blow without hesitation. Mickey met the swing, forcing it low. The usual move from there much the same: lock the limb up and disarm. But Mickey was very much aware of two other players approaching from his rear. And with Angela at their mercy. And was therefore disinclined to use half measures. Other Guy still worrying about taking control of his knife hand back, with Tall Guy's example preoccupying his thoughts, when Mickey went in for the knockout.

He chopped a knife hand into Other Guy's neck. His go-to vagal nerve strike. Dropping Other Guy, already unconscious. His weight pitching him forward in a face plant onto the pavement. Mickey flicked a glance to the other two. Pleasantly surprised to find them unexpectedly tied up dealing with Angela. Giving him time to go back to Tall Guy. Who had dithered between fight and flight a moment too long.

The knife now held awkwardly in his left hand. Mickey dropped, sweep kicking his feet out from under him. Dumping him on his coccyx, probably hard enough to break it. Sprang back to his feet, wrapped up the knife hand and broke his left wrist, just for good measure. Letting the blade drop to the pavement. Took him by the hair and dragged him up into a sitting position.

'Stay here. I see you trying to leave, I'll use that knife to hamstring you.'

Got a white-faced nod of the head in reply. Tall Guy having gone from would-be killer to hapless victim in the space of thirty seconds. His face pale grey with shock. A marginal semi-pro at best. Now coping with the sudden drop back to amateur status. Mickey reached into his pocket and took his phone. Did the same with Other Guy. Then walked swiftly back to Angela. Whose two guys were both leaning on the wall. One of them cursing volubly. Both rubbing at streaming eyes.

'Mace?'

'Yes. You feel like letting them off?'

Mickey considered the two men. Eyes streaming, throats wheezing. Bunched his fists. 'No. Not really.'

49

The police were on the scene within two minutes of Angela's call. Angela taking charge, pushing Mickey into the background. Instructing the skipper in charge to check out her MI5 status. Awed coppers surveying the human wreckage while their skipper got busy on his Airwave. With nothing to do other than help the broken would-be assassins into their carriers. Resistance not simply futile, but impossible, given the physical damage Mickey had done to them. The inspector arrived three minutes later. Having received very clear orders from Met Control. That Ms Stewart's instructions were to be followed to the letter.

Angela's instructions being that the would-be assassins were to be arrested. And kept incommunicado for twenty-four hours. Stressing that last instruction. Citing the Counterterrorism and Security Act. Telling the inspector that he absolutely had the right to delay his prisoners' phone calls. After which the Met could do whatever they wanted with them. Which, since neither she nor Mickey would be giving a statement, would have to be their release. The inspector nodding equably. And instructing his troops to get them into their carriers and away to the nearest Accident & Emergency.

One officer per arrestee. No phone calls, no talking, no nothing.

'What were you, before you left the Job?'

Angela smiled wanly at Mickey. 'I was wondering if you were ever going to ask. Happy enough if you didn't. I was a DI working out of West End Central. Homicide and Major Crime.'

Mickey nodding. Too knackered, post-combat, to say anything. Just grinned. Quietly proud of the way she'd gone about dealing with the other two. And reassessing her capabilities on the fly.

They met James in an all-night café a suitable distance from the restaurant. Mickey having called him from Angela's phone. Having already put his own phone in a handy flower bed just in case it was being passively tracked.

James walked in. Saw Angela and looked slightly guilty. Sat down opposite Mickey and picked up the tea waiting for him with a nod of thanks.

'I swear I had no idea there were other Service officers in the house.'

Mickey nodded. 'It's not a problem for me. If Box wants to keep its assets in the dark then it can't complain if they end up forming relationships that otherwise might be seen as inappropriate.'

A momentary pause was the closest James came to surprise. 'Ah. I see. Well, your point is well made. But a little irrelevant. You're both withdrawn as of this afternoon. So whatever you might choose to do in your free time is—'

'No.'

'No? No as in you're going to disobey a direct instruction and continue with your task?'

'Yes.'

James looked at the table for a moment. 'I realise that you were close to Mr Smalls. And I completely understand the urge to visit retribution on his killers. But you cannot simply ignore the instruction to withdraw.'

'Watch me.'

'If you go within half a mile of Upper Grosvenor Street you'll be arrested. Charged with something meaningless and kept in a cell for the next seventy-two hours. Is that what you want? Why not just go home and wait this thing out? Leave it to the Service and the Regiment?'

Mickey leaned forward. 'Because, for one thing, I can't go home. Kuzmich tried to have me killed half an hour ago. If I go back to a known address there's every chance he'll try again. And on top of that, you need me in there tomorrow morning.'

James, seemingly unperturbed by the news of the attack on them, thought for a moment.

'So they're onto you then. We can worry about how later. Even more reason for you not to go anywhere near the place. And what would be the point anyway? What good can putting you at risk do?'

'You need someone right at the heart of whatever it is they're doing. Because whatever you think is happening tomorrow, I think you're underestimating them. Because if they've escalated to killing MI5 officers then they're playing for the jackpot. Why take that risk unless the pot's there for the taking.'

Mickey raised a hand to forestall James's reply. 'And there's something else. Obviously I was in close proximity to Salagin during the race weekend. And two or three times I caught him looking around with a sort of wistful expression. Like he was saying goodbye. Plus Kuzmich met him when we landed, looking like he had things to tell him. I think there's more going on tomorrow than you suspect. And you need me in the middle of it. Because when I'm proved right, you're going to need to know where they are. Aren't you?'

50

Abir was late. Not deliberately so. His lateness was the result of an unexpected call to join his fellow jihadis. The other four men who had sworn to have Allah's vengeance on the *kaffir* who persecuted their brothers across the length and breadth of the Prophet's lands.

The message had come in as he was preparing for bed. A knock at the door, and a child waiting when he opened it. With a slip of paper that bore a single line of text.

'Join your brothers in the usual place.'

Impressive operational discipline, and Abir was well placed to judge such things. Being, in reality, not a would-be martyr, but an MI5 officer. University-educated, then recruited by the Service on graduation. With a career path that was intended to take him past the street level, but which made clear that the street level was where he was needed in the first instance.

Provided with a bulletproof legend, including a second-generation Pakistani family in Bradford who would happily attest to his being their cousin. Their revenge on the extremists for the loss of their son to a bungled explosives plot. Leaving them with nothing to bury and a hole in their lives that demanded payback.

Working slowly and patiently, Abir had managed to ease his way into the heart of the community. His story that he had run away from a godless home in another town. And had, little by little, been accepted as the fanatically religious man he purported to be. Eager to make the *kaffir* pay for their crimes against the brotherhood. Three years of slow, steady progress resulting in induction into the heart of the struggle.

All the time with his ear to the ground. Making contact with his case officer only when he had something to tell that might make a difference. And doing so via snail mail. No electronic traces to be picked up by the suspicious or simply astute. Abir was nothing more than a hole in the water, to borrow a metaphor. Invisible.

He hurried to the meeting. Cursing the child for having left it so late to deliver the note. And so was distracted when he knocked at the door twice, waited and then knocked twice more. It opened, and he hurried in, already apologising for being late.

Not realising that he *was* the meeting until the bolts were shot behind him. And he saw the camera set up and waiting. The floor covered in plastic sheeting. And the long knife laid out on a cushion. And his so-called brothers, waiting for him with their faces set hard. Along with a stranger. A man whose face was scarred, and with the hard, bright eyes of a killer.

'Good news, brother.'

Their leader, kissing the stunned agent on both cheeks. Abir allowing him to make the greeting even as his mind reeled at the scene's implications.

'Tomorrow morning we are called to the service of Allah.

To give our lives in battle, and to slay the unbelievers in their thousands.'

He waited, raising an eyebrow at Abir as if to invite a reply. Who fought a stutter he hadn't suffered for years to make some sort of answer.

'That is... indeed... good news.'

'Is it not? But there is also less happy news. Delivered to us by the same comrades in this holy war who have called us to battle. It seems that one of us is not what he seems. Not a brother in the unceasing struggle to bring Allah to the godless. He is a spy for the *kaffir*, we are told. And we are instructed to make him harmless before we travel south.'

Abir was unable to reply. And was looking about him, seeking some way to escape when they came forward and pinioned him. Tied his arms behind his back and knelt him in front of the camera. Their leader bent to speak in his ear, gesturing to the knife. Which Abir saw the hard-faced stranger lift from its resting place. Vanishing from view as the unknown man walked behind him and took a grip of his hair to pull his head back.

'We lucky few are unlikely ever to return to his house. The godless will kill us, in due course, when we have reaped as many of them as we have time for. And as holy martyrs we will receive the privileges guaranteed by Allah. Forgiveness will be ours, and a seat in paradise, and the jewels of belief. We will be assured of favour, on the day of judgement, and crowned with the crown of dignity, a ruby of which is better than this whole world and its entire content. Each of us will be wedded to seventy-two of the pure dark-eyed houris, and our intercessions on the behalf of seventy of each of our relatives will be accepted. We will be blessed a thousand times over.'

He stood and walked away.

'Whereas you, Abir, and all the *kaffir* you conspire with, will have a very different experience.'

He signalled to the brother behind the camera. Who triggered the recording and watched, rapt, as the knife made its first stroke.

51

The ghosts left their hide for the last time at 0700. Ivan the last man out. Waiting for Filip to climb out of the access tunnel into the cottage's kitchen.

'All set?'

'The warhead is armed. Set to trigger three hundred seconds after the pressure pad in the tunnel is activated.'

The sixth *Solntsepyok*. Powerful enough to destroy all evidence of the hide. And ready to incinerate anyone trying to enter the empty facility. Orders from Moscow, along with the detailed plan for Target Zero. To protect the hide from discovery. Ivan followed his demolition expert out to the car. Pausing to check his appearance in the cottage's hall mirror. Looking sharp.

All four of them had clippered their hair down to stubble. Shaved off their beards to present a clean impression. And dressed in uniforms intended to inspire trust. Then loaded the fifth warhead into the vehicle that would be both disguise and transport. Ready for the day of reckoning.

52

Mickey walked into the Upper Grosvenor Street house as if nothing had happened the previous night. Ten minutes before the start of his shift. A study in relaxed normality. Put his topcoat into a tray. Pushed it into the X-ray machine's tunnel. Winked at Angela and walked through the metal detector. Raised his voice and called out to Dimitri.

'Hey, Phone Man! Come and get it!'

The Russian put his head around the doorframe. A rare sighting outside his natural habitat. Hand out, waiting for Mickey's phone. Mickey made him wait a moment, then spread his empty hands wide.

'No phone today. I lost it in a fight last night. Rough old city, isn't it?' Paused a beat, then spoke again as the hacker was turning away. 'Any recommendations for the replacement? Given you were always looking at my old phone like it was a piece of shit?'

Dimitri shrugged. His response dismissive. His opinion of Mickey's technical capability clearly low. Happily allowing Mickey to distract his attention from the screens if it allowed him to demonstrate alpha-geek supremacy.

'An old Nokia, perhaps? Given all a man your age will do is make calls and send texts?'

Mickey grinned. Soaking up the piss-take. Knowing Dimitri had given his location to the thugs the previous night. Exchanged a meaningful glance with Angela. Who had already expressed a strong preference for the punishment due to Dimitri. And hearing the hiss of rollers as his coat left the X-ray. Angela having seen exactly what he had up his sleeve. But not Dimitri. His back to the screen. And clearly with no idea that Angela was MI5.

'Fair enough. A new iPhone it is then.'

Picked up the jacket and made his way to the front entrance. Managing to keep his face straight when Kuzmich rose to greet him. Barely controlling the urge to go to work on the Russian. And take vengeance for Kev. The Russian's face unreadable. Although the assassins' silence would clearly have been reported back to him.

'Michael Bale. All is good with you?'

'Been better, been worse, Mr Kuzmich. And yourself?'

'I'm fine, thank you for asking.'

The Russian was armed. The butt of something heavy duty peeping out from beneath the carefully cut jacket. Mickey guessing at Heckler & Koch. Hand cannons for the discerning penis compensator.

'You're carrying? Is that necessary?'

Kuzmich shrugged. 'Pavel Ivanovich has a meeting in the city. It is, shall we say, sensitive. And so I will accompany him. You will drive.'

Mickey nodded. 'Shall I warm the car up?'

'No need.' Kuzmich smiled. A dangerous expression,

Mickey judged. Feeling more than ready to meet the threat implied. 'It is already at operating temperature.'

Both men turned to the lift as it opened. Salagin stepping out. He greeted Mickey with the usual polite formality. All three men riding the lift down into the basement. Where the Merc, true to Kuzmich's word, was waiting. Engine ticking over, hazing the air with carbon monoxide.

Mickey opened the nearside passenger door for the oligarch. Closed it for him, folded his coat and went to put it in the boot. Finding a pair of overnight bags already in situ. Went round to the driver's seat and got in without comment. Noting that Kuzmich was lounging in his seat in the usual manner. Without the seat belt fastened. Eased the car onto the vehicle lift and waited while it was hoisted to street level. All three men silent. As the garage door opened Mickey was the first to speak.

'The City, Mr Salagin?'

'Yes please Michael. Harbour Street.'

GazNeft. Mickey drove on autopilot. Noting without comment the car that pulled in close behind them. Kuzmich's pointed glance back at the occupants a clear indication that its presence wasn't unexpected. Leaving Mickey to drive uninterrupted. Wondering what business Salagin had with the Watchmaker on the day of a Sunburn strike. And why a second car was deemed necessary.

He eased the S Class through the mid-morning traffic to their destination. Parked in the usual traffic warden magnet spot. Salagin murmuring that he wouldn't be long and to wait for him there.

Mickey, bracing himself for a long and uncomfortable silence. And so was just a little surprised when Kuzmich leaned towards him. Guessing what the Russian intended even as the other man lifted the armrest. And happy to let it play out. Watching as he thumbed the print reader and lifted the stubby little automatic out. Unsurprised to find himself looking down the barrel's small dark snout.

'Has nobody told you that it's unwise to point firearms, Yuri Sergeivich? On top of which, isn't taking a second gun just a little greedy?'

Kuzmich chuckled softly. 'You are not in a Bond film, Mr Bale. No amount of amusing dialogue is going to alter the situation you find yourself in.'

'The situation I find myself in?'

The Russian's smile broadened. 'Maximum points for cool. And for having a pair of balls on you. I have to admit I was very surprised to see you this morning. I would have thought you might have taken the hint by now?'

'What, when you paid a bunch of amateurs to try to take me out?'

Kuzmich shrugged. 'I was thinking of the way your colleague was so easily dealt with on his secret motorcycle. But that too. However it is that you made them vanish, I confess myself impressed. My associate had every expectation that his men would deal with you effortlessly.'

'It wasn't that hard. They were dependent on their knives. The same way you're hanging onto that Glock. Like a baby with his teddy bear. And what makes you think I won't do the same to you?'

The Russian shook his head. 'The fact that I will shoot you if you so much as twitch, perhaps?'

Mickey shrugged. Happy to go on letting him think the weapon was any sort of deterrent. 'OK, so you have me helpless. To what end? I mean you either have to kill me or let me go. At which point I will call the police and have you all arrested. Should look good in the papers. Billionaire racing team owner jailed for illegal firearms possession. Sidekick gets twice as long just for being a thick bastard.'

Kuzmich's expression hardened. 'You have two choices. You can die. Unlamented and unmissed. Or you can swallow all that bile and escape with nothing more damaged than your pride.'

'You forgot option three, which is where I kill you with my bare hands.' Mickey grinned. 'But let's ignore that for the time being and assume you're right. Why would you let me live? Last night you were paying men to knife me to death.'

The Russian shrugged. 'I thought you had a sense of humour, Michael Bale. They failed. And in defeating them you earned enough respect to merit being spared, as I see it. And besides, Pavel Ivanovich has a soft spot for you. But live or die, it really makes no difference to me. We will be gone inside the hour. Gone forever. To somewhere the British state will never be able to touch us. You are wondering why the second car? It is insurance. The plane that is waiting for us has a take-off slot in precisely fifty minutes. And nothing, not an accident, not a breakdown, is going to stop us being on it.'

'Ah.' Mickey nodded, as the confirmatory penny dropped. 'You've given the Sunburn crew their final target. And now it's time for you to run away before they carry out some sort of 9/11 attack. What is it then? The Shard? The Changing of the Guard?'

'Closer to home than either. Even now they are approaching their target. And not just the *Prizraki*, the ghost soldiers who have eluded your hunt. We've also used your extensive databases of extremists to bring all them together in a way that won't be forgotten for a hundred years. First the Sunburn, and then a hundred jihadis will run amok through your city.'

Mickey digested that. 'So why is Salagin in there?'

He gestured to the GazNeft office.

'He is signing legal control of his entire commercial empire over to the Russian state. He'll keep the trappings of wealth, of course, as the nominal chairman, but the *Rodina* is claiming back what was only ever on loan. Let's face it, once all this is traced back to him, as it will be, the Magnitsky Act would have been used to take it all off him in any case. Better it goes back to the country it belongs to, yes?'

'Clever. If the link to you is discovered the Russian state will let him take the blame. Side-stepping any formal accusation. And presumably sheltering him from prosecution. I can hear the disavowals already. How does Pavel Ivanovich feel about being a target for the rest of his life?'

Kuzmich shrugged. 'When you have enough money, the idea of retribution becomes a difficult concept to take seriously. Even when there's a country as capable as yours involved. Plus we'll be a very long way from here, in a country that is no friend to the UK. And Pavel Ivanovich is a loyal son of the *Rodina*.'

Mickey, wondering if Salagin had always seen it that way. Or had had to be reeled in, a little at a time.

'Of course we knew that MI5 would be sniffing around after us. Which made your true purpose within Pavel Ivanovich's household obvious from the day you arrived.'

Kuzmich's facial expression had become smug. But then he probably thought he could afford a little gloating, Mickey decided. Given he was looking at Mickey down the barrel of the G26.

'We would have known exactly what you were even without the gambling.'

'Gambling? What…'

He made the connection. Kuzmich seeing it in his change of expression.

'Yes. The leak that ruined Pavel Ivanovich's plan to recruit a famous driver as the result of much gambling on the matter. We asked Moscow to find out where it was centred. And database infiltration revealed that the majority of the gamblers lived close to a town called Hereford. The link was not hard to deduce. Given you were the only person whose presence could have drawn your Special Air Service to the meeting.'

Mickey decided that saying nothing was likely his best choice. Denial probably useless. And agreement unwise, given he was probably being recorded.

'Nothing to say, Michael Bale?'

He shrugged. 'Nothing beyond fuck you.' And smiled back at Kuzmich. Thinking I know something you don't. And waiting for the Russian to come at him across the centre armrest. Likely to swing the pistol at his head. Use it like a knuckleduster. Likely to result in his getting it back in his face. Leaving Mickey with a host of options thereafter. But to neither his disappointment nor relief, the only response was a matching smile.

'I have to admit that I have a sneaking respect for you. You take risks. You are a dangerous man, if allowed room to operate. Much like myself. In truth I no longer wish to kill you. And there is no need to take the risk. You can drive us to the airport. We have private parking nearby, under cover. You will be restrained, and placed in the trunk. To be found in due course, I imagine. If you make enough noise.'

Mickey raised an eyebrow. 'You're going to let me drive? Aren't you worried I might put the car into a lamp post?'

Kuzmich shook his head. 'Not really. For one thing I have an airbag. And a second car. And for another, if you even twitch in the wrong direction I'll unload this weapon into your legs. I'd imagine that'll take a while to kill you. We'll be out, into the second car and away inside a minute. You'll be stuck in that seat, bleeding to death. Do it my way and you will live. Do you really want to spend your life without good reason?'

Mickey put on a show of thinking it through.

'No. Not really.'

'Very wise.'

The two men waited in silence. Kuzmich putting the easily concealable pistol down by his side and smiling at a traffic warden. Winding the window down and accepting the ticket. Mickey pondering shouting for help, but deciding not to do so. Mainly to spare the innocent's life. But also because there was no real need. Kuzmich pursing his lips and nodding. As if saluting Mickey's pragmatism. And gloating over his victory.

53

'You called an immediate and urgent, Susan?'

The Deputy DG, framed by the operations room's doorway. Susan Miles gestured for him to come into the room. Having pressed the panic button to summon him fifteen minutes earlier. Her prerogative, although not one to be exercised without very good cause. No other option, in this case. As it become clear that something unprecedented was happening.

'We thought we had one jihadist cell heading into London. All under surveillance and all with a Regiment strike team in close attendance. That just changed.'

Harding looked at the map. Showing eleven locations. All streaming live video, running in picture-in-picture windows.

'What platforms are they streaming on?'

'Facebook Live. All of them. Video of themselves making peaceful protests. With non-inflammatory placards. The locations on the wall screen's map encircling the city. From Ealing Broadway to Stratford Westfield, Hampstead High Street to Lordship Lane in Dulwich. Twelve potential massacres, given the men involved. Facial recognition popping up alerts identifying a dozen of them as ISIS foot soldiers. Men last seen in Syria or on the migrant trail to Europe. Known killers. Police scrambling to respond.'

'We didn't know this before now?'

Susan maintained her professional demeanour with admirable composure, James thought. Given the hint of disbelief in Harding's voice.

'There was no warning. All of these cells have been quiet verging on dark for weeks. Someone must have instructed them to stay quiet. Someone who knows everything about them.'

Leaving unstated the obvious fact. That the information regarding their locations could only have come from within Thames House.

James had realised something was going on as the level of noise around the web analysts' desks grew. Members of the intel team quickly repurposing to join them as the number of WhatsApp and Telegram messages between the cells went ballistic. An obvious shout of 'Hi, here we are!' to the authorities. And then the Facebook videos had started popping up. One after another. Designed to grab their attention.

'And no sign of any violence?'

Miles shook her head.

'No naked blades. No firearms. Nothing that looks remotely like it could contain a bomb and no-one bulky enough to be wearing a bomb-vest.'

The DDG stared at the screen thoughtfully. 'So this is either some new and altogether more subtle tactic, or something much worse. What's the status of the Met's response?'

'We've got Tactical Support Group and MO19 teams in transit from their usual locations. ETAs between two and nine minutes. Answering calls for urgent assistance from local officers who've got potential bloodbaths on their hands. Or thought they did, when they called it in.'

'Which means that their usual locations are suddenly lacking the usual level of protection. Is it possible that we're being had?'

54

The ghosts were on station. Tucked away down a side street close to the river. Hidden in plain sight. Less than a hundred metres from Target Zero.

Ivan, sitting in the back of the vehicle, looked down at the warhead beside him. Carefully decorated with decals intended to disguise its purpose. Or at least reassure the people among whom it would be planted. Sure, it still looked like a rocket warhead. If you knew what a rocket warhead looked like.

But to the innocent it was just a two-metre-long tube. The same diameter as an oversized drainpipe. With the word 'POLICE' in blue on yellow. The exact Pantone references used by the capital's police. So when the time came to place it, why would anyone worry? Especially as it would be carried by four uniformed coppers.

Until it was too late.

55

After another ten minutes Salagin reappeared from the building. His face expressionless as he got into the car.

'It is done. Time for us to leave.'

Kuzmich gestured with the pistol. 'City Airport. You know the way.'

Mickey nodded. Put the car into drive and whispered it away from the kerb. Checked that his intended trajectory across the pavement was clear. And, knowing that he had just enough time before the planets aligned, indulged himself. Flicked a smile at Kuzmich. 'Let's see how good your stoppage drills are.'

He floored the accelerator. Turning the wheel to aim for one of the skyscraper's steel supports. The steel leg pretty much cosmetic, of course. The building's weight mostly resting on the central core. But it was still three feet thick. Engineered to support thousands of tonnes of the building's outer construction.

Kuzmich acted without hesitation. Raised the pistol. Put it within a foot of Mickey's head. And pulled the trigger. Rewarded by nothing more than a click. The hammer falling on a bullet whose primer failed to fire.

To be fair, Mickey decided later, the Russian's reactions were pretty much spot on. Racking the weapon's slide as the car accelerated hard towards its final resting place. The useless round spinning in the air as he ejected it. The slide snapping forward to collect the next round. Kuzmich's finger squeezing the trigger again, in the instant before collision. And then he was in flight.

The limo hit an immovable steel pillar at fifty miles per hour. Speed built up in less than a hundred-metre flat-out sprint. Thanks, Mickey knew, to Salagin's insistence on always having the most powerful engine available. Six hundred and twenty horsepower more than enough to do the job. Bringing the Merc's intricate automotive machinery together with the building's construction-strength metalwork at twice the urban speed limit.

Mickey rode out the impact, restrained by his seat belt. Further assisted by the deployment of the driver's airbag. Unlike Kuzmich. Who, lacking any restraint from a belt, found himself airborne. Probably expecting the passenger airbag to arrest his progress towards the windscreen. But if so, disappointed in that expectation.

Mickey had recognised Kuzmich's threat early. A threat reinforced by his insistence in sitting alongside Mickey when he was driving Salagin. And had realised that he was at his most vulnerable behind the wheel. A realisation that had led him to do two things in the hope of mitigating the risk he posed.

The first made possible by Errol. Who had provided the 9mm rounds. Dummies, from his display case. Lacking both a genuine primer and the propellant explosive to fire the bullet. But looking genuine in every respect. Two of which

Mickey, under the cover of cleaning the car after an outing, had inserted into the G26's magazine. Making sure that the first trigger squeeze and the one that followed a swift rack of the slide would both fail. Rendering the weapon useless.

The second, ironically, the product of his renewed friendship with the Friday Night Boys. And Steve's new job managing security for a car dealer. Steve having prevailed on Mickey to come and say hi to the dealership's top technician one evening. Sammy. Ex-Job, and keen to shake Mickey's hand. Having worked the same relief fifteen years before. Mickey with a very specific question in mind. A question the answer to which he had swapped for a few minutes discussing the gunfight in the Fulham High Street. Sammy nodding and excusing himself for a minute. Popping into the dealership's parts store. Then leading Mickey to a big limo just like the one Salagin owned.

'We're not allowed to do this, of course. I could get sacked for just showing you how it works.' He'd winked, pulling a small piece of electronics from his pocket. Like an elongated stripy bead with a wire poking out of each end. 'See this resistor? It's all you need. That and knowing which wire to cut.'

He'd bent down into the rear passenger foot well. Pulling the wiring harness free from beneath the front seat.

'See that wire?' Pulling the wire in question free and pointing at it. 'If you were to cut it, and then connect this resistor to the two ends of the wire, the airbag won't work. Because that's the wire that tells the car there's someone sitting on the pressure pad. If that's cut, even if the resistor makes the car think it's still intact, then the airbag isn't activated.'

And so Mickey had established a routine of carefully

cleaning out the footwells with every return to the garage. Making it easy to snip the wire in question and tape in the resistor.

The result being that Kuzmich's forward momentum was unhindered by the airbag. Resulting in his impacting the windscreen. Hard. Crazing the glass. And knocking him senseless in the same instant. Mickey pushed the airbag away and unclipped his seat belt. Leaned over and retrieved the compact Glock from the foot well. Racked the slide to clear out the second dummy round. And nudged Kuzmich with the muzzle. The Russian lifting his head as if with a great effort. Staring glassily at Mickey. Uncomprehending. 'Wha…'

'This is for Kev.'

He squeezed the trigger. Putting a bullet into the Russian's head. Kuzmich's corpse lolling back against the passenger door. Then turned to face Salagin. Who was extricating himself from the rear passenger airbag's abrupt embrace. Staring in horror at the spray of blood from his dead comrade's violent demise.

'The dead bastard said you had a soft spot for me. And I've got one for you.' He pointed the gun at the oligarch's foot. 'There it is.'

'You can't—'

Mickey shot him. Putting the bullet into the top of his immaculate Church's Chelsea boot. The oligarch's face going pale as the shock and pain hit him. Mickey hit the button that opened the boot lid. Reasoning that Salagin would be needed alive, so using the metal sheet to protect the rear window.

He reached over and unclipped the snap on Kuzmich's shoulder holster. Pulling the pistol free. Having guessed correctly. H&K P30L. The 'Wick Stick', beloved of action film aficionados. And the harder-punching version; .40 Smith & Wesson load; 10mm, rather than the usual 9. Got out of the car, crouching low. Catching the follow car's crew dismounting less than twenty feet away. Waited to see how they planned to play the situation. Unwilling to start shooting if all they intended was rescue.

He spotted the snout of a machine pistol momentarily pointing upward. The driver stepping round his door and bringing the weapon to bear. Enough firepower to shred Mickey in half a second, on full auto. Mickey beat him to the punch. Took aim, centre mass. Then shot the Glock's magazine dry in one swift volley. Slapping the gunman back against his vehicle. Swapping the .40 into his right hand as he hurried, bent double, around the Merc's front. One on one with the passenger. The other guy with the advantage of a pocket machine gun. Mickey with the advantage of a decade of close-quarter combat drills.

He popped his head up for an instant. Pulling it back as the other guy spotted him from behind his car's door. And fired a long burst. Starring the Merc's glass and hammering the body panels. Mickey made an instant decision and shouted as if in pain. Tossing the empty Glock into the open. Hoping the other man would see the weapon and assume he'd got lucky.

He waited two seconds and then popped up with the H&K ready. Finding the other man rising from cover. Presumably having taken the bait. Mickey fired a double tap. Putting both bullets clean through the glass window and into his chest. And the gunman went down and stayed down.

Priorities. He opened the passenger door. Put the pistol's muzzle to Salagin's undamaged foot.

'What's the target.'

The oligarch shook his head. White-faced with the shock, but with his jaw set grimly. 'I cannot tell you.'

'And it's only a foot, right? I get it.'

He put the pistol's muzzle to Salagin's crotch.

'If I pull the trigger now you'll be without your manhood for the next forty years. Whereas if you co-operate, you might even get the chance to use it inside the next decade. Whatever you decide, make it quick. Because the one place you're not going is back to Russia. And when your boys set that warhead off, the only person left alive to take the blame is going to be you!'

Salagin closed his eyes for a moment. Opened them again. 'You can't keep me safe.'

'No, but I can testify that you tried to do the right thing.'

The Russian nodded. Opening his phone out into tablet format. Stabbing at the screen with a trembling finger. 'Here.'

Mickey looked at the picture on screen. A map of central London with annotated arrows highlighting a dozen points. 'Jesus...'

He picked up the empty Glock. Grabbed the spare magazine from the armrest. Unloaded the .40 and tossed it into the car. Too much of a giveaway, given its size. Got his coat out of the Merc's open boot. And walked away fast. Knowing that running would make him stand out. Took a screen shot of the map and saved it to a text. Sent it to James's mobile. And vanished into the side streets.

56

The operations room was at the centre of a slow-motion nervous breakdown. With a direct line to the Met's control room. Which was fielding urgent requests for heavyweight back-up from local reliefs. Some with two or three groups on their ground. All shouting for help to deal with a potential London Bridge. Local coppers anxiously getting ready to wade in with their truncheons.

Tom's field teams were continuing to track the original target cell. Assumed to have been heading to a meeting with the presumed Spetsnaz team. Increasing bafflement at their actions creasing Susan Miles's previously serene features. Incomprehension growing, as they walked slowly west along the A501. Heading for Regent's Park. Which, obviously, was no-one's choice of a prime terrorist target.

'This starts to have the feeling of a disaster.' Harding. No judgement in his tone. Just simple fact. 'The group we expected to be the main threat seems to have gone passive. And now we have a dozen unexpected pop-up threats. So widely spread that it's taking all the special-duty assets the Met can scramble to get them under control. Someone's been very clever indeed, putting all this together. And even if we do pin them down, we still don't know where the Sunburn is. Which means that the SAS teams just have to wait in place until something happens. Something presumably being the bloody thing being detonated somewhere busy.'

And in the chaos, the fact that officers from Kennington police station were responding to a 999 call went unnoticed. A vehicle collision outside the GazNeft Tower. Shots fired. Completely under the radar for Operation Resolve. And James's phone, set to silent, went unnoticed when it vibrated to announce the arrival of Mickey's message.

57

Mickey threaded his way through the maze of streets to the east of Waterloo Station. Considering the plan laid bare by the schematic on Salagin's phone. Looked at his watch: 11.42. Curtain up on the morning's spectacular at 12.00 on the dot. And the plan, he was forced to admit, almost perfect.

First, an explosion. A Sunburn warhead, capable of destruction not seen in London since 1945. When the Nazi *wunderwaffen* urban redesign scheme had fizzled out. An atrocity to be swiftly followed by a focused jihadist attack on the survivors. An attack that would qualify as an outrage even without the bomb. The carefully planned run-up to the strike rendering the Russians' target zone defenceless. As the Met scrambled to get back control of an unfolding situation. A dozen known terrorist groups suddenly live-streaming from suburban high streets around the capital. A dozen potential bloodbaths. Each one needing to be contained. Dozens of officers being sent to face down each potential flashpoint.

Mickey looked at his phone. Still no response from James. He copied the plan/map again and pasted it into another text. Sent it to James with a terse message. 'READ THIS NOW. ATTACK AT 1200. MOVING TO INTERVENE AT BOMB SITE.' Then repeated the action with Angela as the recipient.

Looked at his watch again: 10.44. Sixteen minutes. Put the phone in his pocket and started running.

58

Angela felt her phone vibrate in her jacket pocket. Having ignored the house rules and kept it on her person. Knowing she'd be no use to Mickey if it was locked away.

'Toilet break, Andy. Back in five.'

The other guard nodded, deep in his book. Barely looking up as she walked away. Stopping in the corridor, she opened the text. Stared at it for a moment. Initially disbelieving. Disbelief promptly replaced by the realisation that Mickey was in deep shit. Made her mind up. Took the lift down to the bottom floor. Walked through the pool area and kitchen with a swift, purposeful stride. Waving a hand at Chef's greeting. Knowing she'd never see the woman again.

She took the baton from her belt. Standard issue inside UGS. Kuzmich wanting his guards to be able to put up a fight, if necessary. Walked down the corridor that led to staff reception with a soft tread. Looked round the corner of Dimitri's office door.

The hacker was staring at his screens in apparent consternation. Sensing her presence he turned, nodding distractedly as she smiled at him. Only realising at the last minute that she wasn't really in a friendly mood.

With a flick of her wrist Angela extended the baton. Spun in a tight 360 to build the required momentum. Extending her arm to give the baton's iron ball tip as much power as possible. Battering him into the screens with an extravagant backhand blow. The hacker lay across the keyboard, semi-conscious. Giving Angela a moment to consider the map. Both cars immobile outside the GazNeft building.

She pondered that while she was flexi-cuffing Dimitri to his chair. Gagged him with her uniform scarf. Nice and tight, so that he could barely breathe through his mouth, let alone call for help. Then stood back to consider her handiwork.

He mumbled something through the gag. Angela taking a moment to work out what he'd just said. What he was going to do to her when he got free. She smiled and shook her head.

'Two things, sonny. One, you're in no position to be making threats. Are you?' She tapped him on the temple with the baton's tip. Hard, making him yelp through the gag. 'And second, you sound like you're the sort of guy who likes the idea of rape. Which, from a woman's perspective, makes you in need of some mental readjustment. Let's start with the basics. What you just promised me was sexual violence, right?'

He stared at her sullenly. No less pissed off and desirous of revenge. Just keeping his mouth shut.

'Thing is, Dimitri, sexual violence isn't *always* good for the guy. Let me demonstrate.'

She hammered the baton down into his crotch with enough force to do serious damage. Kind of hoping that she had. The response was immediate. A thin screech that died away to frenzied sobbing.

'There you go. Sexual violence, not enjoyable. Just hold

that thought for the next fifty years and we'll all be a lot happier. When I say we, of course, I mean women in general. Give the SAS my regards when they kick the door in.'

She pushed the chair over to prevent him wheeling it to the door. Left him lying on the carpet and closed his door. Slapping on a Post-it with *'Back in 5'* scribbled on it and headed for the exit.

59

James's phone vibrated in his trouser pocket. Three swift pulses. Prompting him to pull it out far enough to see the sender's ID. A number, unfamiliar. Social engineering, marketing, whatever. To be ignored.

'Do you have a view, James?'

Susan. Staring across the table at him like she'd caught him cutting his eraser up in class. He pushed the phone back into his pocket.

'I'm sorry, Susan, do I have a view on...?'

'We were discussing whether this sudden rash of jihadists popping up could be part of something larger.'

James nodded. 'I'd say it's very possible. After all, they seem to have managed to wake up all these cells. Why wouldn't they be able to bring the rest to the party too, just not so high-profile?' Realisation dawned on him. 'How many active jihadist cells are there, that we're aware of?'

'Seventeen.' Katy, her eyes hard with the realisation of what he was implying. 'We know of seventeen cells. But there are only twelve of them broadcasting right now.'

'Which means that there are five more out there

somewhere.' James looked around the table. Seeing the same realisation sinking in.

'The plan was never just to detonate a warhead in London. They want to tie it to our own home-grown terrorists. And have the Met at full stretch coping with a dozen potential London Bridge-style attacks all over the suburbs. Allowing the remaining cells to wreak bloody havoc, uninterrupted by anyone with a firearm. Which will not only have everyone thinking it's an ISIS hit, but it'll also take the attention off their operators while they get out of the country. Along with whoever's been supporting them. And I'd bet good money that were we to raid GazNeft right now we'd find them clean. They'll have been getting ready for this for weeks.'

Harding shook his head.

'We can't even do anything to stop them, because we don't know where the target is. We could quite conceivably be looking at another 9/11 here.'

60

Belvedere Road. The closest point to the target accessible by road. And Mickey made the bomb delivery vehicle immediately. Having to doff his cap, mentally, at the sheer bravado involved.

A fake police car. Parked on the double yellows outside the office block facing the target. Like anyone was going to move *them* on. And as fakes went, as close to the real thing as to make it indistinguishable. Unless, of course, you were a copper. Or a recent ejectee from the Job.

And not just any old police car. A BMW X5 in silver. MO19 standard. Wearing the Met's tribal colours. The usual tasteful blue and hi-vis dayglo yellow wrap. 'POLICE' written down the sides. The Met's website address on the rear doors. Light bar across the roof looking like the real thing too. Only one thing jarring by its absence. Whoever Salagin's people had got to wrap the car had clearly had access to the actual design. Probably computer-generated, so just a graphics file. Which had included the words 'ANPR FITTED' on the rear quarter light. Except the boxy automatic number plate reader itself was missing from its usual perch on the X5's roof.

The impression of a Met MO19 Armed Response Vehicle otherwise perfect.

Mickey kept walking. Slowing his pace. Ostentatiously looking at the Black Bay. Channelling 'early for an appointment and hanging around' into his body language. Sent another text to James. 'READ PREVIOUS MESSAGE. GHOSTS ARE IN POSITION CORNER OF BELVEDERE RD & CHICHELEY ST. SUNBURN IN FAKE ARV BMW. INTERVENING.'

He spotted a smoker, twenty metres from the BMW. Strolled over to him with a broad grin. Pulled out his wallet and peeled off a tenner.

'Hey, buddy, I *so* need a fag. Gave it up yesterday and it's grating hard, you know? Give you this for a couple of yours and a light?'

The smoker obliged. Amused at a fellow addict's failure to quit. Eager to score a fresh pack for the price of two cigarettes. Lit Mickey's first smoke and wandered back into his workplace. Leaving Mickey with the perfect cover. A suited and booted businessman having a pre-meeting drag. He wandered aimlessly up the pavement, surveilling the fake ARV with fleeting sideways glances. The cigarette burning in his left hand. Right hand in his trouser pocket.

The BMW four up. Two in the front visible, looking studiedly relaxed. Both mid-thirties. Both clean-shaven. Both hard-faced. Looking a lot like soldiers to Mickey. But like coppers to anyone else. Bored coppers, waiting for a shout. Two in the back, dim outlines behind the tints. Probably scanning their vehicle's surroundings with hard, professional eyes. As to whether the last Sunburn was on board, hard to say. But likely, if they were staying on schedule.

The plan on Salagin's phone saying that they were to debus at 1155. Have the Sunburn in place by 1158. Placed at the point where the cables that held the London Eye in position converged on their anchor points. Back in the car at 1159. Make a swift exit and detonate by command trigger at 1200 precisely. A detonation that would be audible for miles. An unmissable signal to every jihadist who hadn't yet declared their presence as part of the deception. The call to start the mayhem.

Mickey glanced over at the Eye. Pretending to enjoy the last of his first cigarette. The queue to board probably five hundred strong. Most of whom would be inside the warhead's lethal radius. Perhaps the wheel would be blown from its mountings into the river. Perhaps not. A valuable publicity gain for Moscow if it were. An image to resonate down the decades. But either way the mass slaughter would hit tourism like a nuclear weapon going off. Not to mention the deaths in the subsequent extremist bloodbath.

Millions of visitors would choose to give London a miss. Deciding they'd be safer in Rome, or Paris. Not to mention the lost revenue from people deciding not to go out after all. Putting the hospitality sector back onto the bones of its already seriously underfed backside. As body shots between countries went, pretty effective.

He looked at the Black Bay. Just after 1153. Time up. Support or no support, Mickey had to act. Had to take on four Spetsnaz operators and win. Or die. Along with hundreds of innocents.

61

Ivan flicked a glance at his watch.

'Two minutes. Situation check.'

All four men looked out of the car. Alert for any sign of danger to their mission.

'Sasha?'

Sasha checked the right of the vehicle. 'Nothing to report.'

'Anton?'

The front. 'Nothing to report.'

'Filip?'

Left side. 'Nothing to report.'

Ivan twisted in his seat. Scanning the area behind the BMW. Nothing. Just a middle-aged guy with a cigarette in his mouth. Unremarkable. Nondescript even.

'Nothing to report. Very well, we—'

The pedestrian rapped at the passenger near-side window. Making the universal gesture asking for a light. A stupid apologetic smile plastered across his face. And an unlit cigarette between his lips.

'Hi, guys! Anyone got a light?'

62

Mickey had strolled around the corner. Putting himself momentarily out of sight from the fake ARV. Put a hand into his coat pocket. And taken out the device that Dimitri had missed earlier. The hacker distracted from his screen first by Mickey and then Angela. Having made it ready, he'd put it back in the pocket. Firmly gripped by his right hand. Then went back round the corner, flicking the first cigarette's butt away ostentatiously. Getting into character.

Then he'd shaken his head angrily. 'Realising' that he didn't have a light for the second smoke. Pretended to think for a moment. Then turned to the BMW. Putting the second cigarette between his lips. Knocked on the passenger window. Miming the use of a lighter with his left hand. Raising his voice to be heard through the glass. Grinned around the smoke's butt. Doing his best to look like the daft twat he wanted them to see.

'Hi, guys! Anyone got a light?'

Knowing he was flat-out gambling. Betting that the Russians would find his request reasonable, if unwelcome. Betting that someone would lower a window and offer him a light. Or to tell him to do one. Not caring, as long as the window opened. And betting that they wouldn't see his right hand, hidden in the coat pocket, and go defensive. Or worse, loud.

The passenger turned stone-like eyes on him, beneath a Met-issue firearm officer's baseball cap. METROPOLITAN POLICE and a chequered band. Eyes that Mickey realised had seen more than their fair share of death and destruction. If only from the calculation in the stare. He repeated the mime. Refusing to go away. Another moment's delay. The two front-seaters talking. And Mickey getting ready to smash a window if he had to. Then the passenger reached forward. Pushed the switch that lowered the sheet of glass. The window opening a few inches. Mickey gauged the gap. Enough.

'Fuck off.'

The accent thick. The instruction accompanied with a glare. Mickey's hand already in motion. His body turned away to put his right arm behind him, as seen by the passenger. As he lifted the hand out of the pocket. Pivoting, thrusting his right arm out. Posting the cylindrical grenade through the gap. Just making it as the glass started to slide up again. The lever springing free as it fell onto the passenger's lap. Cuing pandemonium inside the car. Doors cracking open, the Russians getting ready to dismount. Reaching for weapons. But too slow, given the weapon's short burn fuse. Way too slow.

Mickey ducked into the cover of the passenger-side door. Leaning his weight against it as the man behind it frantically tried to escape. Shouting something incoherent to his comrades. And the grenade exploded.

Even outside the car the bang was brutally loud. Accompanied by the BMW's windows crazing. The space behind them suddenly filled with white gas; 170dB of prompt hearing loss followed by the sudden inability to breathe.

It was James who'd spotted it in Errol's collection of illegal firearms. Standing out among several other drab painted fist-sized cylinders. He'd reached out and picked it from the shelf. Holding it up to the light to better examine its slate-grey body and muted orange warning stripes.

'Do you know what this is?'

Errol had given the grenade a cursory glance. Busy extolling the virtues of a genuine Russian-made AK74 to Mickey.

'CS gas. That's what the bloke who sold it me said.'

James had nodded, his face suddenly expressionless. 'Put it on Mr Ward's bill then. We might have some use for it, with what we're planning.'

The grenade had been sold to Errol, it transpired, by a member of the Regiment. Along with a grab bag of other military goodies. A mixture of captured kit and items appropriated from Regiment stocks. James recognising the weapon instantly from his own service. An experiment, conducted on operations in Afghanistan. One that hadn't been deemed worth persevering with.

It was a product of Project INLAD. Immediate and Non-Lethal Area Denial. A Ministry of Defence trial weapon. Based on the GLI-F4, a contentious grenade used by the French Compagnies Républicaines de Sécurité. The infamous CRS. Riot police, among other functions. Who had experimented with the recipe for a while in the early 2000s. Before deciding that it was a little too spicy for domestic consumption.

One half of its ingredients being ten grams of chlorobenzalmalononitrile. CS for short. Most police forces' incapacitant of choice. Exposure to which was guaranteed to render the victim temporarily blind due to copious tears. Hence the term 'tear gas'. Accompanied by a burning irritation of the mucous membranes of the nose, mouth and throat. Causing profuse coughing, nasal mucus discharge, disorientation, and difficulty breathing. Result: incapacitation.

The other half of the mix being twenty-five grams of trinitrotoluene. Better known as TNT. The fuse tailored to burn for two seconds, rather than the usual four. Making sure there was no way it could be thrown back. Producing one hell of a bang. And filling the average-sized room with CS in the blink of an eye. Issued to Special Forces units in the Middle East for field trials.

And just as swiftly withdrawn. The Special Forces soldiers in question declaring it to be a bit shit. Of no more utility than a standard stun grenade. And more confusing, in the chaos of an operation. Although it was suspected that not all were returned. Including, apparently, the one sold to Errol. Enough punch to disorient and then temporarily disable the occupants of the average-sized room. Its effect within the confines of a car, even one with the doors open, devastating.

Mickey took a deep breath. Pulled the BMW's passenger door wide. Allowing the stricken fake copper to fall out onto the pavement. His face peppered with bloody wounds where the grenade's casing had spat micro-shrapnel at him. His slitted eyes streaming bloody tears. Barely able to breathe, blinded. Mickey assessed him. Threat: none. The automatic in his belt holster tempting, but too loud. Left him to it. Taking the firearm cap from him and putting it on his own head. Immediately camouflaged as a copper. And just as well. Over the top of the car he could see people in the queue for the Eye staring across at the wrecked BMW. And all it would take would be couple of off-duty US Marines to put a crimp in his next sixty seconds.

The left-side rear-seater was out of the car. Swaying on his feet from the blast's concussion and hanging on to the door with one hand. Reaching for his pistol with the other. The go-to reflex of the gun dependent. Mickey put a straightforward half-fist into his throat, dropping him choking to the flagstones, then moved on.

Walked quickly round the back of the car to find the driver already out. And holding a carbine. Standard Met issue MCX. The GRU really getting their homework right. But unable to find a target through eyes streaming from the effects of the tear gas. Mickey stepped inside the weapon's muzzle sweep and pushed the barrel aside. Delivered a brutal punch to the side of the Russian's neck. Dropping the man, already unconscious.

And then ignoring him. As the BMW's right rear-seater put the muzzle of a silenced Glock in his face. The man behind it running on instinct. Levelling the long barrel at the only target he could make out through his streaming tears. And pulling the trigger.

63

Operation Revoke's team leads were still trying to work out where the Sunburn was going to be detonated when Katy from intelligence's phone buzzed. Not the usual single beep of an incoming message. Rather, the three rapid and angry pips of a priority update from the ops room. Whose data-gathering agent software was designed to flow relevant material to the right people in real time. She looked down and then read the message out loud.

'Salagin's been found in a car outside the GazNeft building. He's been wounded, gunshot to the foot, and his associate Kuzmich is dead. Head shot. And there are two more bodies, also both shot.'

While that was sinking in, James's phone buzzed again. Someone clearly keen to talk to him. He fished out the phone. Number not recognised. Shit. He opened the message and read it.

'It's from Bale.'

He opened the previous message. And looked at the plan attached.

'Ye gods. It's the London Eye. With jihadist attacks on the Southbank and Westminster Bridge immediately afterwards.

Detonation in…' he shot a look at the clock '…seven minutes. After which the groups live-casting will start attacking civilians. Bale's on the spot and says he's about to take on the ghosts.'

'On his own?'

Harding, with a look of disbelief.

James stood up. 'He doesn't have much choice, does he? The nearest ARV is probably ten minutes away. This is that moment of his disposability you told me about.'

He slid his phone across the table to Susan. 'That's their plan. You need to get the Met to that fake police car on Belvedere Road, by Southbank. Whether he manages to deal with them or not, that's where the bomb is. And *don't* call him back – he'll be busy.'

He turned to Harding. 'Call the east entrance and have the police on duty ready to accompany me. Tell them I need a weapon!'

He ran across the building. Dodging staffers, ignoring their shouts of surprise. Took the stairs three at a time. Making a beeline for the east entrance. Opening onto Milbank, next to the river. Where he knew there'd be the usual pair of armed policemen. Who were waiting for him by the main doors. Both of them looking nervous.

'Major Cavendish? Your boss just phoned my boss; we're seconded to you.'

One of them looking fit, and handy with it. A constable. His sergeant, who had greeted James, at least ten years older. And looking like he was carrying two stone of surplus weight.

'Give me your weapon, Sergeant. I need you to stay here and keep the door secure with your sidearm.' A face-saving way of telling the guy he wouldn't keep up with them. Not that the man in question looked anything but relieved. 'Radio your control and tell them that Major Cavendish and this officer are heading for Parliament Square. I want armed officers waiting for me there in two minutes. Tell them there's an immediate and urgent threat to life. You...' he turned to the younger of the pair '...with me!'

64

Mickey dropped to one knee and both hands. Ducking under the pistol's threat. The Russian's first shot flying harmlessly over his head. Reckoning afterwards that the silencer had made it just that little bit unwieldy. A fraction slower to aim.

He pivoted on his hands. Throwing out a lunging sweep kick. Upending the Spetsnaz trooper as he lowered the pistol to reacquire his dimly discerned target. Pounced on the fallen man before he had the chance to regain the initiative. Grasping his right fist with his left hand and smashing his right elbow into the Russian's face. Not the perfect attack. A little weaker than he'd have liked. No major muscle group behind it. But better something than nothing, as Chuck had drummed into him throughout his training.

'You got the other guy hurting, right? What you gonna do, give him a few seconds to get his shit together? This is either MMA or the street, Mickey boy, and either way you put that asshole down and then you keep him down. You extemporise. You attack, you attack, you fucking ATTACK!' Chuck's eyes blazing as he spat the words out. 'Fists, knees, elbows, head, whatever the fuck works. Just do something! You beat on that *motherfucker until he isn't thinking about fighting anymore. No mercy, you hear me?'*

Still on one knee, he swayed back a little to regain his balance. Contemplating his options. Seeing the Russian still grasping the pistol. Lethal determination on his scarred face. Figured two, perhaps three seconds to reorient and send an unhealthy dose of 9mm Mickey's way. Realised he had no time to get up. And extemporised.

He reached out and gripped the Russian's throat with one hand. Pushed him back against the open car door. Blocking the pistol with his left hand. The Spetsnaz taking the bait. Fighting him for control of the all-important weapon. Ignoring the hand at his throat. Perhaps figuring he could fight for a moment or so without breathing. Big mistake. Mickey made his fingers into claws. Tightened his grip on the other man's windpipe. Digging in to get his fingers around it. Then squeezing mercilessly. Hard enough to crush it shut. Perhaps irreparably.

The Russian, realising his peril, loosened his grasp on the pistol. Allowing Mickey to twist the wrist away and strip the gun from his fingers. Putting the weapon's barrel to the other man's leg and pulling the trigger. Fight over. The Russian arched with the pain. Temporarily losing all interest in resistance.

Mickey stood, sized him up and took his hand. Lifted it from the wound and triggered a bullet through the palm. Try holding a gun with that. Did the same to the unconscious driver. Walked around the back of the car and repeated the trick on the other back-seater. Feeling a little queasy but knowing he had to render them both helpless and easily identifiable. Ignored the front-seat passenger, blinded and still fighting to breathe.

Feeling his eyes starting to prickle, he took a breath and put his head into the car. Slitting his eyes to avoid as much of the CS gas still floating in the air as possible. The warhead was resting in a frame, protruding from the load area between the two rear seats. Smartly decaled in sliver and dayglo yellow, to match the car. The word 'POLICE' in blue on a yellow background. A nice touch. Nothing to see here. Police business, move along. Until it exploded. What looked like a remote detonator lying on the back seat. He pocketed the controller. Detonation averted.

'Police! Stay where you are!'

The voice went quiet as Mickey turned, silenced pistol in hand. A lone copper. And young enough to be a probationer. Truncheon drawn and looking determined. Poor bastard. What a shout to have answered. Raised his left hand. Ready to put a bullet into the kid's leg with the gun in his right if he had to.

'Before you try have a go and make me stop you, I'm ex Met. Prot. And current MI5. Nothing to prove it, except these four aren't Job, they're Russian. And that's a Sunburn warhead in the back, right? So you can be the hero, stand guard on this lot until the cavalry arrive, right? But be ready to get on the ground when they do, given this lot look like MO19. Oh, and you'd better take these.'

He dropped the silenced pistol's mag. Racked the slide to eject the chambered round. Then offered the kid the weapon. Of no further use to him. What Mickey needed now was something that would work as advertised. Something familiar.

And with a lot more noise. To shock and awe the men he was about to go after. The young copper reached out a hand.

'No, put a pen through the trigger guard. Don't get your prints on it, they'll obscure what's already there. Don't they teach you lot anything these days? Oh, and there's this.'

Handing over the remote detonator.

'This is the on button for the bomb in the car. Don't push any buttons or you're liable to flash-fry yourself and a couple of hundred tourists too. Just give it to the grown-ups when they get here.'

He took the Glock 26 from his jacket pocket. Thinking here we fucking go again.

'What are you doing?'

He smiled wanly at the baby copper. 'What you lot would be doing, if you hadn't been dragged all over London by these clever bastards. Trying to protect the innocent.'

And then the screaming started.

65

James and the AFO were panting as they ran up the western side of the Houses of Parliament. James wishing he'd done a bit more exercise in the past few weeks. He stopped at the Parliament Square gates, breathing hard. A pair of armed coppers from Parliamentary and Diplomatic Protection coming out with their weapons half-raised. Lowered when one of them recognised their mate beside James. Shirt sleeve order and flak vests, their carbines loaded and ready. Looking uncertain.

'I'm Cavendish, MI5! There are armed jihadis on the bridge and they're about to start massacring innocents. You're with us! Follow me!'

He led the three armed policemen around the corner and onto Bridge Street. Past Westminster Tube on the other side. Carrying his weapon down by his side, barrel pointing at the ground. Deliberately non-threatening to avoid spooking the hundreds of civilians milling about on the bridge approach. And hopefully to avoid providing the men about to unleash hell upon them with any warning. He raised his voice to be heard by the firearms officer. Knowing that they were about to undertake a life-changing experience.

'Make sure your weapons are set to single shot. Aimed shots only. Aim centre mass, keep shooting until they go down. And try to angle the shots down if you can. Things could get bad enough without us slotting civilians with misses and shoot-throughs.'

They passed the pelican crossing at the junction of Bridge Street and Victoria Embankment. Still seeing no sign of any would-be terrorists.

'If this goes off the way it's supposed—'

He raised a hand to halt their advance. Then took a knee, ducking into the cover of the black rubber traffic barriers blocking the pavement. Patted the air, signalling them to do the same. Fighting the urge to giggle as the people walking past them divided into two camps.

Those people determined to ignore them. Locals mostly. Apparently used to coppers doing copper things.

The other half being the unashamed gawkers. Tourists for the most part. One young woman even stopping to take a photo. Blocking the walkway, hemmed in by the steel barriers protecting the pavements from passing traffic.

'Move on please, miss?'

She stared blankly at the PaDP constable for a moment, then did as he had asked. Having taken a photo that James suspected might just be about to make her wealthy. The moment before the shooting started.

'Listen in.' He waited until he had their attention. 'Two men, black hoodies and jubba jackets. Fifty metres out, with their backs to us. Probably more of them on the other side of the bridge.'

Not needing to explain the likely tactic when it went to the races. The two groups would use the threat of their knives.

Herd the people between them into an increasingly confined space. Cutting throats and inflicting life-threatening wounds. Or just throwing their victims into the river to drown. And all on live stream, most likely. He checked his watch: 1159.

'There might be a bloody great big bang over the river very soon. The signal for them to start the slaughter. It might not happen, in which case they'll be delayed a moment or two. But either way those men have to be taken down. Hard. I'm going over there.' He pointed to the other side of the road. 'Because there are probably more of them on that side.'

He took off his jacket and wrapped it around the carbine. Impromptu camouflage that needed to work for just a few seconds to prevent the jihadists spotting him. And starting their deadly, short-lived reign of terror too soon.

'One more thing.' He looked at them each in turn. 'All the training you've done was about avoiding hitting civilians. And I need you to forget that. If you get the shot, take the shot. Without hesitation.'

'But—'

The youngest of the three. Still an idealist, James guessed.

'You *take* the shot. And yes, you might get some collateral. You might put the round through the target and into some poor bastard out for a stroll. But if you wait for the perfect shot the target might kill two or three more times before you can drop him. It's the first thing they teach you in the Regiment. Go centre mass and stop the killing. I'll take full responsibility.'

The policeman nodded. White-faced. Ready to fight, blood concentrating at his core. James nodded.

'Stay in cover until I start shooting. When I go loud, you do the same. Take those two down and then double-time it across the bridge to get the rest of them. And good luck.'

He got to his feet. Climbed over the divider that separated pavement from roadway. Dodged through the slow-moving traffic and hopped over the divider on the other side. Weaving through the foot traffic at a swift walk. Head down and channelling his inner grey man. Spotted the terror cell members in front of him when he was twenty metres from them. Two men. Not as overt as their comrades on the bridge's southern side, but unmistakable.

One of them turned in his direction, clearly making him for security forces immediately. And drew a pistol. Spending precious time adopting the two-handed shooter's stance he'd seen on the TV. Let off a round that flew wide. Mercifully not hitting a civilian. The reaction instant. Some people dropping to the ground. Others running in random directions. And some frozen where they stood.

James had already dropped the jacket. He raised the carbine and put the death dot on the armed man's chest. Took up the slack and squeezed, putting the first round through him dead centre. The jihadist toppling backwards against the bridge wall. He shouted the legally required challenge, switching targets.

'*Armed police!*'

The other man had a knife. Long and sharp. Raising it as the people around him began to react. Slowly, subconsciously unready to recognise the threat. Starting to duck, starting to flinch back, but too slow. James sidestepped a fleeing man dragging a child. The runner's gaze back at the knifeman rather than where he was going. Costing precious seconds as he reacquired the target, momentarily lost behind the big man's bulk.

And found him again. The knife now red. Rising back into

the air from its first stroke. A middle-aged woman staring in disbelief at the blood stain spreading across her chest. James unable to see centre mass, the jihadist surrounded by fleeing pedestrians. Even his head intermittently obscured. Making it the kill shot no better than fifty-fifty. Either hitting the target or killing an innocent. James shrugged, mentally. Knowing he had no choice. Took the shot. The terrorist's head jerking back as a 5.56mm copper-jacketed slug deleted the malign intelligence that was driving him to kill.

A moped whizzed past between traffic and barrier, jinking around a bus into the middle of the road. A woman, unhelmeted, hair streaming in the breeze of her passage. James filing that under non-essential and pushing on.

On the other side of the road the three policemen were firing. The constable looking across at him, carbine still raised but without a target. James waved him forward with an urgent gesture. On the eastern side of the bridge fresh screams were erupting from the terrified crowd.

More terrorists, making the most of their moment of vengeance for all of their imagined and actual grievances. At least one person leaping over the parapet into the river to escape their rage. The traffic stalling as pedestrians hurdled the barrier and spilled into the road. Desperate to escape from the knives that James could see rising and falling behind them.

He started advancing into the fleeing crowd. Shouting 'armed police!' and shouldering aside the panic-stricken who failed to see him in their path. Knowing that people were dying within a hundred metres. Breaking into a run and putting his shoulder down to bull through the press of civilians scrambling past him.

66

Mickey hurried towards the Eye. At least as much as hurrying was possible, pushing through a panic-stricken mob. Moving forward into the open, tree-lined space between the damaged BMW and the river. An expanse of paving that had until a moment before been the scene of an orderly crowd. Waiting for their admission to the barrier-controlled final queue. Ready for what the operators grandiosely called 'boarding'. A quiet Monday lunchtime gathering. Orderly and good-natured.

A crowd that had instantly become a mass of panicking individuals. Some of them already spooked by the sight of Mickey taking down the Spetsnaz team. And now terrified by the screams and gunfire from the bridge over the Thames a hundred metres distant. Flocking blindly away from the threat. Individuals and small family groups following their instincts to survive. Their flight the tide against which Mickey was pushing. Heading in the opposite direction. His pistol and replica MO19 baseball hat parting the flow of bodies to either side of him.

Fifty metres distant, by the riverbank, someone screamed.

Not the swift, breathless call of a parent trying to hurry their children along.

Not the involuntary gasp of fear as a runner was infected with the terror of the people fleeing alongside them.

The last frenzied exhalation of a life being extinguished. Horror, terror and something more existential. Fading away as the person doing the screaming died, Mickey guessed. He pushed through the fleeing tourists to his left. Stepped up onto a three-foot-high stone balustrade. And caught a fleeting glimpse of a pair of knife-wielding men. Dragging a terrified woman between them. An empty tourism information booth's door hanging open.

The Southbank jihadis clearly having heard the shooting on Westminster Bridge. Armed police from the Palace of Westminster, perhaps. And decided to start their brief blood-soaked reign of terror. Knowing they'd be caught on camera. Arrested, if they weren't killed while resisting.

And not caring. Because this was their moment in the light of a crimson sun. Their long-dreamt-of chance to wield their blades against the godless *kaffir*. To scribble the dogma of their perverted version of Islam across London's pavements in blood.

He started to raise the pistol. Realising it was pointless even as his arms twitched. The range too great for the Glock's stubby barrel. Just a waste of precious ammunition. Then flinched as she went down under their frenzied attack. The two men repeatedly stabbing at what was probably already a corpse. Mickey felt his anger rise. The rage uncoiling from deep within him. The actions required to take down the Russians had been calculated. Ice-cold. But Mickey's fury was now a raging furnace. Demanding that no quarter be given. No survivors to stand in a dock and crow at the victims' families. No. Prisoners.

He jumped off the balustrade. Pushed through the last fleeing remnants of the crowd. Raising the Glock, two-handed grip. Cleared the last runners. And found himself in the face of his enemy. Two to his front, twenty feet distant. Two to his right, perhaps thirty. Another three closer to the river. Still butchering what looked like an American family. Dad, already dead, garish check shirt over a blood-stained white T-shirt. Mom, screaming and falling under their knives. Two small children, transfixed with fear. Unable to run.

Ten rounds in the magazine. One in the chamber. He contemplated taking the three by the river first. Sparing the children. But forty metres was still too distant. Not the percentage choice.

He walked forward. Boring in towards the closest two knifemen. Who were done with the dead tourist advice officer. Turning to face him with blood extravagantly painted across their kufi jackets. Ten feet. He shot the closer of the two in the chest. Dead centre, where the organs were clustered. The percentage shot. Then raised his aim to snap a second shot into the blood-spattered man's bearded face. Nine rounds remaining.

Peripheral vision telling him that the men to his right were frozen. Still reacting to the unexpected attack.

The second of the two murderers ran at him. Shouting something in the expectation his god would hear him. And perhaps bring divine intervention down on Mickey. An incongruous climber's axe raised by his right ear just in case Allah was busy. Ready to hack and tear at the unbeliever's flesh.

Mickey waited, adjusting his aim. Shifting his weight slightly onto his left foot. Let the snarling Islamist get to within five feet. Then shot him once, through the forehead. Lights out. Stepped aside as the corpse's momentum carried it past him on its way to the ground. The dead man's beard igniting from the propellant's incandescent kiss. Eight.

If their comrades' deaths deterred the other cell members in any way, it wasn't obvious. The man to Mickey's right, his front now that Mickey had sidestepped left, screamed blue murder. A big man. Raising two knives and roaring for vengeance as he stormed forward. Mickey's first shot slightly high and right. Any small deviation of aim punished disproportionately by the short four-inch barrel.

The bullet punched into his upper chest, between clavicle and shoulder. Seven. The jihadist staggered, but kept coming. Mickey adjusted and fired again. Punching a copper-jacketed slug into his attacker's chest. Dead centre. The big man went down. Six.

Another extremist came out of nowhere from Mickey's right. Having presumably passed him unseen in pursuit of the crowd. Then realised that a wolf was among them. And decided to be the hero, rather than keep running. Bad choice.

Mickey bent his knees, ducking under the wild knife swing. Then rose to face his attacker, so close that he could feel the other man's breath on his cheek. Gripped the knife hand at the wrist, twisted and locked it up. Put the Glock into his lower abdomen and pulled the trigger. The report muffled by the layers of clothing. The bullet going straight through and spraying the paving stones behind him with blood. Five. Lifted the gun and put it under the terrorist's jaw. Looked into the other man's eyes and blew the top of his head off. Four.

The last of the closer jihadis dithered. His righteous fury dispelled by the gale of fear Mickey was bringing. Turned to run. The first round clipped his shoulder; the second hit him square in the back and smacked him to his hands and knees. Two rounds left. He started walking towards the last three men. Who were standing over the corpses of the children. Their clothing and weapons red with the blood of innocents.

Mickey's anger boiled over. Staring at them as he put the Glock's barrel against the stricken jihadi's head. Roaring his rage at them.

'No prisoners!'

He pulled the trigger. One.

One bullet and three enemies left. He put the blade sight on the closest man's body. Centre mass. Then indulged himself and raised the point of aim. Fired, knowing as the pistol bucked his point of aim was perfect. Hit the Islamist in the throat. Dropping him, blood foaming on his lips.

The other two ran.

Mickey tossed the empty weapon away. And sprinted after them. Given wings by the adrenaline flushing his system. Seeing a woman fleeing in front of them both. Too terrified to turn aside into the park to her right. Her fear gave him additional purpose. He caught the closer of them in twenty strides. Took him down hard. A rugby tackle, using the other man's body to cushion his fall. And was back on his feet while the jihadist was still on his hands and knees. Shaking his head in confusion, having smacked his head hard on the paving slabs. Mickey raised his left arm high. Then stepped forward and whipped it down and back. A fast bowler's trick.

Providing the essential momentum for his right arm's hammer fist. Using the corner point of his right little-finger knuckle to deliver the killing blow. A precise strike between shoulders and skull. Aiming for the uppermost of his cerebral vertebrae. Snapping the jihadist's spine.

He looked around for the last of them. And then found himself on the ground. Looking up at the sky. Head ringing. Unable to move.

Something blocked out the sun. Mickey squinted, forcing concussed vision to focus. It was the woman. Unveiled. Wearing make-up. The perfect camouflage. In her hand, a claw hammer. Which made it no surprise that Mickey felt like an anvil had been dropped on him.

The last of her comrades joined her. Looking grimly down at Mickey, knife held ready to use. And Mickey no more capable of resisting than of jumping across the river. The woman touched her fellow terrorist's arm. Both of them ignoring the growing wail of sirens in favour of revenge. She dropped the hammer and took the knife from him.

'Let me do it.'

She bent over Mickey, sizing him up. Working out where to put the blade in. Raised the knife, picked her spot and tensed to stab the point down. And then she was gone. The thud and squeal only dawning on Mickey's lagging consciousness a moment later. He raised his head groggily, seeing a blue uniformed figure battering the last jihadist with an extended baton. The terrorist already reeling from what he assumed was a head shot. Angela? She attacked again, and the terrorist went down choking. His larynx ruptured by a scything blow that had connected with his throat. Fell to his knees, scrabbled at the pavement as he realised that breathing was no longer possible. Unable to protect himself from the follow-up delivered across the back of his neck. Then lay still.

Mickey struggled to his feet. Looked down at the female extremist. Dead. Her face contorted. Blue-hued, a red weal across her throat.

'You weren't taking prisoners either?'

Angela shook her head. 'I saw what they did to those kids.'

She put the baton on the ground and got down on her knees. Signalling for him to do the same as the first uniformed figures came sprinting along the wide walkway. Carbines raised, eyes wide at the slaughter.

'Armed police! Get down on the ground!'

They lowered themselves onto the paving slabs. Mickey realising that getting up was going to be far more difficult than lying down.

67

The police quickly decided that Mickey wasn't in any fit state to give a statement. Still groggy and with a splitting headache. And, in point of fact, hadn't really needed him to do so with any urgency either. Given the trail of dead jihadists he'd left them to analyse. The incident commander, informed by James just who Mickey was, had him put in an ambulance. Angela insisting on accompanying him.

Five hours later, he was discharged by an overworked Accident & Emergency doctor and sent on his way. Having been X-rayed and run through an acute concussion assessment. Deemed fit to go home, under supervision. Angela promptly delegating herself that task.

James picked them up outside St Thomas's. Bundled them into the back of the Service taxi. Mickey shaking his head, then grimacing at the pain. Momentarily looking for Kev in the driver's seat. Then remembering that Kev was dead. James installed himself on one of the tip-up bulkhead seats. Grabbed a strap as the taxi pulled away.

'You do realise that you were exceptionally lucky?'

Mickey narrowed his eyes in thought.

'How do you make that out?'

'He means, idiot...' Angela, affectionately chiding the slow child '...that you took a full-on smack with a hammer and all you've got is a bad headache.'

Mickey nodded. Angela stroking his hand with no sign of letting go. Having clearly decided that she was keeping this one.

'Apparently it was partly her size and partly her lack of technique. Not forgetting the hat.'

Mickey shook his head at James again. 'The hat. What, does it have some clever Kevlar lining?'

'She was only five foot one. Which meant she was too short to get a proper strike on you. She hit you low down, behind your ear, and the hat band cushioned the blow.'

'Still felt like being hit by a train.'

'I'm sure it did. But there's no fracture. Not even a hairline. Whereas a proper blow would quite probably have pushed a piece of your skull into your brain. Resulting in the medics having to trepan you, put a plate in and so on. Lucky boy.'

'He's not the only one who was lucky, is he, Mr Cavendish? I saw you take that head shot. I was the one on the moped.'

James grimaced. Having managed to forget the risk he'd taken, shooting through a crush of fleeing tourists to kill the first jihadist he'd taken down.

'I simply played the percentages.'

Angela nodded. 'So, forgetting about Medal Magnet here, what happened on the bridge?'

'We were lucky. The Parliamentary and Diplomatic Protection guys I took with me did most of the hard work. One of them had the presence of mind to get up on the bridge parapet. Used the extra height let him see his targets. There were eight of them all told, and he took down five. I'd imagine he'll be shaking a royal hand very shortly.'

'How many people died?'

James winced. Having walked back down Southbank and across Westminster Bridge once he'd established that Mickey was still alive. The roadway empty, other than for the last of the medics and the first of the forensics teams.

'Seven around the Eye, not counting enemy casualties. Mr Bale was on them too quickly for them to get into their stride, so to speak. Another eleven on the bridge. Probably twenty or so more with various degrees of wounding.'

He fell silent for a moment, contemplating his memories of the scattered corpses and blood-soaked survivors.

'And we were lucky with the other cells too. The Russians had their timings out just enough to make the difference. If they'd ordered them to trigger their live feeds five minutes later then they could have killed dozens of civilians before the Met could get to them. Instead of which the TSG and MO19 rounded most of them up before they had chance to do anything. And should be able to mop up every one of them, when the few runners are tracked to earth.'

'Is that what we call a result then?'

James looked at Mickey for a moment. Pondering the desolation in his voice.

'I understand, Michael. Right now it feels like failure. But when the awful pain in your head goes away you'll be able to reflect on the five hundred or so lives you probably saved. We've analysed the plan from Kuzmich's phone, and we reckon that the Sunburn would have killed about two hundred people. Double that if it had actually knocked the Eye off its mounting. Add in the probable casualties on the bridge, and the slaughter they'd have wreaked on the Southbank among the survivors of the blast, it's not hard to see this rivalling 9/11 for impact. Your country is very, very grateful to you.'

'Grateful enough to be slinging GCs around?'

James smiled wanly at Angela.

'Grateful enough for it to be gongs all round, I'd imagine. I've already been offered a bar to my George Cross, and I expect Michael will be too. Plus I think you can expect something to be bestowed upon you, Ms Stewart. Albeit in private, of course. Investitures for operational Service personnel are usually quiet affairs. After all, I can't see the DDG wanting to let golden boy there go anywhere anytime soon.'

Mickey shot him a disgusted look. 'So this temporary blackmail is set fair to stick, is it?'

James smiled bleakly. 'It's the first rule of intelligence work at the sharp end, apparently. If you're not good enough you tend not to last. And if you are good enough they never want to let you go.'

68

'It's a fucking shame. He was one of a kind, Albie.'

Sammy Chin nodded. His expression the model of solemn sincerity as he replied to his colleague. Although nobody at the table was fooled for a second.

'They broke the mould when he fell out of it, that's for sure.'

General agreement around the table. Insincere, but completely unanimous. Albie's fellow gang leaders having gathered at the Bosphorus Café. The shared neutral ground of their very informal association. To celebrate his life. Or, to be more accurate, his death. All parties generally being of the opinion that Albie had been a pain in the arse. Too full of his own importance. Too much the old-school Brick Top emulator. While the current zeitgeist was for understatement. And for the sort of feigned restraint that would make a Japanese yakuza clan boss nod approvingly. Albie had been too flashy for everyone else's good. His death regrettable only in terms of its inauspicious timing.

The plan was for a memorial breakfast. With champagne. After all, they were celebrating, after a fashion. Followed by a trip out to the house. A swift visit before the swiftly organised funeral. Low-key, high ostentation. Proper brass handles and solid oak. Ostensibly a collective paying of their respects.

In truth, verification. Making sure that it was actually Albie who went into the ground. It being in nobody's interests for the old bastard to do an Elvis. His grieving relatives having already been gently warned. Display of his body prior to burial a condition of their inheriting what was being allowed by the families. To allow them to make the appropriate gestures of respect, of course.

'Problem is, it means there's another territorial division to be worked through. And so soon after Joe's ground had to be allocated.'

Sammy nodded again. The thought having been very much on his mind. The problem being that Albie's ground had bordered Joe's. Which had made him a prime claimant after Joe's death. Now his enlarged empire was itself up for grabs. And needed to be divided up quickly, before the rats started gnawing at it. He opened his mouth to share his thoughts on the matter, then closed it again.

'What the fuck…?'

His attention had been drawn by a flurry of activity out front. Visible through the Bosphorus's big front windows. Police uniforms suddenly swarming the pavement. And no ordinary coppers. Certainly not local. Territorial Support Group, at a well-informed guess.

The Met's hard cases. Often ex-military. The blunt instrument specifically designed to deal with anything physical beyond the ability of a local nick's relief. Only the lack of facial tattoos differentiating them from a Maori tribe going to war. When the rhythmic drumming of batons on riot shields started, most sensible rioters knew it was time to opt out. Smartish.

A warrior tribe, the TSG. Motivated by their tribal blood oaths to leave no man standing, if it went to the races. Fire-retardant overalls rather than the usual black trousers and white shirt. Not really necessary on the High Street. But, Sammy guessed, sending a nice clear message as to who they were. The men and women around the table watched them with a degree of jealous admiration. Their men standing guard outside the café shepherded meekly away. Desirous of neither being nicked nor a swift and deniable spot of rough and tumble. Allowing themselves to be temporarily detained in the carriers that had disgorged the stormtroopers. And leaving only the individual minders gathered at their own table in the window. Who, it had to be said, looked more than a little nervous.

The café's door opened, and two men entered. Both of them suited and booted. Black ties. Either showing respect or a Blues Brothers tribute act. One of them held the door open. Looked at the minders and asked a simple question. The other strolling across to order with a smile for Sebnem behind the counter. The man at the door exquisitely well spoken. But looking more than capable of going straight through anyone who got in his way.

'Would you gentlemen be so good as to give us a few minutes' privacy? Innocent ears, and all that.'

Not the sort of language or accent usually heard around the manor. Polite, clipped and reeking of eau de posh bastard. Raising eyebrows at the bosses' table. And occasioning the minders, perplexed, to look to the bosses for a decision. Along with the speaker. Serenely confident in his well-cut suit and black necktie. And, it seemed, totally untroubled in their august presence.

'I suggest we agree to this request.'

Sammy, flicking a finger at the door. Not feeling the need to point out the presence of several TSG officers on the other side of the window. The minders got up and left. And were ushered away down the pavement and out of sight. While the other man smiled across the counter at the astounded proprietor.

'Two cups of tea and two of your famous bacon rolls each please, Mrs Bakirci? To go, I think.'

Sebnem inclining her head in acknowledgement of the respect shown. Making it less likely that she'd spit in the tea. And the newcomers walked across to the bosses' table. The posh one doing the talking.

'Ladies and gentlemen, I would be grateful for a moment of your time. If, that is, you can see your way to giving me your undivided attention? I see there's an empty chair.'

He sat in the chair left vacant to represent Albie's memory and leaned back, looking around the table. Not waiting for their agreement. While his comrade stood behind him. Pondering the gang leaders one at a time with a keen interest.

'You are of course wondering who I am. What sort of person, you're pondering, can call the Assistant Commissioner Operations and have four carriers full of permanently angry unarmed combat enthusiasts at his command an hour later? And the answer is simple. My chain of command, while a little arcane, ends in Downing Street. I represent a branch of the UK government that would rather not be identified. Not that you've never heard of it, of course. Everybody has.

It's just that the men who control its operations are a little publicity-shy when it comes to individual cases.'

He shrugged equably.

'Unsurprisingly, I suppose, given it's *secret*. And, I feel constrained to add, they react very badly to breaches of their operational security. So, for the purposes of discussion, we'll just call them the men who'll have you all jailed if you fail to co-operate. For a long, *long* time.'

Sammy suddenly felt grateful for Albie's absence. Reckoning Albie would have been up and in the posh bloke's face in a heartbeat. Occasioning, Sammy suspected, arrest. Detention. And a pestilence upon all their houses. Because it didn't take a genius to work out what was happening. Although the why was still eluding him.

'You might want to cut to the chase, Mr...? Given Sebnem's going to have your order ready in a minute or so.'

The newcomer smiled faintly. 'Oh, I don't think names are necessary on my part, Mr Chin. I think I'll let my credentials do the talking.' He gestured to the man standing behind him. 'Mr Bale, of course, you already know. And I'm sure you also know he used to be a part of the organisation whose collars I'm currently holding.'

He waved a hand at the TSG officers waiting outside. All staring in like they fancied a nice bacon roll. Either that or the chance to pile in on the collected criminal royalty on the other side of the glass.

'Now, to cut to the chase as suggested, here's what I need from you. Attempts have been made, of late, to blackmail a member of the public. And to force him to perform acts of violence on—'

Sammy, being a bright boy, made the connection instantly. Looking round at the man standing behind him.

'The Mantles, for one. Right? I hear you did an outstanding job on those animals.'

He fell silent as the stranger raised an admonishing finger.

'One more word, Mr Chin, and I think we'll have *you* taken away and charged with a variety of felonious misdemeanours.'

Sammy kept his mouth shut. Reckoning the posh geezer meant every word.

'Thank you. Now, where was I? Oh yes. While the initial blackmail attempt might have gained the desired result. I don't think Mr Ward quite appreciated what he was getting into. Which meant that he was the unfortunate subject of what we'll just call compensation for all the discipline he clearly didn't get in his youth.'

'That was...'

'Yes. It was. Who else do you think dispenses correction with explosives? Anyway, Mr Bale here is, as of now, utterly off limits. If he gets even the hint of a follow, or sees someone he doesn't like the look of, I'll have two choices. Well, three, but inaction doesn't really square with me, so let's call it two, shall we? I'll make one of two phone calls. To the Assistant Commissioner Frontline Policing, perhaps, asking him to move in for all of you at four the next morning. Or perhaps, if I'm feeling grumpy, the call that just kills you all.'

'There's no *way* that you could get away with—'

Mickey smiled as James raised a hand.

'As an old drill sergeant used to say to me a long time ago...' He slapped the table. Hard. '*Wait for it!*'

And smiled, happy with the collective start back from

the table that introducing a touch of the parade ground had engendered. Just to remind them what they were dealing with.

'I didn't say *we* were going to kill you. Your own people will do that.'

'How d'you make that out?'

Cigarette husky, one of the two women at the table. Focusing the intensity she'd used to kill her way to the top of her little tree. James smiled more broadly. Knowing that Mickey would have loved to deliver his next line.

'Because, *Martina Cole*, that call will be to my former colleagues in the 22nd Special Air Service Regiment.' Eyes narrowed around the table. James quietly recognising to himself that while he felt dirty doing it, he now had their complete attention. 'Or rather to a little-known operations cell within the Regiment.'

'What, the SAS has got an anti-crime unit?'

Disbelieving. And refusing to back down. The female boss sneering back at him.

'No, more like an abduction team. Men... and women... who make people like you disappear.' Brought his hands together, then flicked them open simultaneously. 'Pouf.'

And stared levelly at her. When she spoke again her bravado was less evident.

'You said our own people would be the ones doing the killing.'

'And they will. All we have to do is hold you incommunicado for a month or two. Treating you well enough, of course. We're not sadists. Just realists. We'll keep you from talking to anyone for long enough that a new leadership will establish itself in your fascinatingly Darwinian little world. At which point we'll release you in the middle of what used to be your territory. I'm sure you can work it out.'

Silence fell as every man and woman around the table pondered that fate. And the likelihood of a new leader happily handing the reins back.

'And in the meantime we'll be monitoring your activities by all sorts of high-tech methods. To see if you're behaving. And to be sure we know where you are if I feel constrained to use my nuclear option.'

All of which was, of course, utter bullshit, as James knew well enough. There was no covert disappearance capability in Hereford. Nor any will to one. And no way Box would waste a minute on surveilling these lowlifes. But the threat was a powerful one. Made very real by his forbidding presence at their table.

'And you can consider yourselves lucky. You especially, Mr Chin. Mr Ward left us with no choice but to terminate his rather sordid existence. Or rather to offer him a choice that we suspected, quite correctly, he was bound to get wrong. Whereas you have not yet quite strayed over that line. Please don't?'

Fresh subdued consternation broke out around the table. James looked over at Sebnem, who was in the act of placing the bagged bacon rolls on the counter.

'Ah, breakfast! Do carry on with the funeral rites, one and all!' James stood. Then turned back to the table. 'You might find Mr Ward looking a little... unnatural, when you see his body. It's a side effect of death caused by overly close proximity to a grenade when he pulled the pin on it. Just thought you ought to know. Right, come along, Michael, let's go and eat our breakfast.'

'We'd appreciate a word with Mr Bale. If that's all right with you?' Sammy. Raising open hands to demonstrate a lack of any threat.

James looked at Mickey. Who shrugged. Curious, truth be told, to hear what it was that Sammy had to say.

James nodded. 'I'll be outside. Don't take too long. Those bacon rolls smell too good to stay uneaten long.'

Mickey sat down in the freshly vacated chair. 'What can I do for you, Mr Chin?'

Sammy smiled back at him. Knowingly. 'Michael Bale. You're a fucking legend, you know that? First off, you're the man who offed Joe Castagna. And now Albie Ward. There isn't a gang boss in the country wouldn't put you on a five-hundred-k-a-year retainer just to have you at his side. Not that I'm offering...' He raised his hands again. Looking around the table to reassure one and all. 'But you've got a problem, Mickey. Not with me. And not with any of us.'

Mickey raised an eyebrow. 'Go one then, shock me. Who's this problem with?'

Sammy leaned forward. Speaking quietly, for effect. 'Your problem, Mickey Bale, is with *you*. MI5? Secret Service? Bollocks. You can take the boy out of Monken Park, but you're never going to take Monken Park out of the boy, eh? And it's worse than that. You've been over the line.'

'The line?'

Mickey, outwardly amused. Inwardly pondering the gangster's words. Sammy adopting a patient attitude. Speaking with a precision and restraint that would have had the hypothetical yak crime lord inclining his head in studied respect.

'There is a line, Michael, between the average man or

woman in the street and the likes of us. We abandon the restraint that is forced upon most people by society's rules. We do as we wish. We steal, extort, sell illegal goods and exploit the innocent. And then we threaten and bribe our way out of any punishment. And you have crossed that line. You belong here, with us. And you know this to be true.'

Mickey pursed his lips. Truth be told, not completely able to gainsay in his own mind the assertion put before him.

'The day that I join any of you will be the day that hell freezes over.'

Sammy shrugged. 'Your reticence is... understandable. After all, the journey from where you used to be to where you are now is not an easy one. You justify the change to yourself by considering the good you have done. Whereas we know that all you have done is shuffled the deck. The cards have moved around, but the game remains the same. And one of these days you will realise that the game is now in your blood.'

He gestured to the window. 'And now, I think you'd best be gone. See you around, Michael Bale.'

Mickey got up. 'Not if I see you first.'

'Don't worry. You won't.'

The gang leaders watched as the ex-policeman walked out. Took his roll and tea from the posh bloke and walked away with him. Leaving the TSG to free their doormen and leave them to the funeral rite.

'You'll never get that one into your firm.'

Sammy shrugged at the opinion. Knowing that the unspoken corollary was that he'd better not try either. Bale looking like the sort of first-strike weapon they all hoped had gone out of fashion when he'd killed Joe Castagna.

'I was just fucking with him. But there was truth in what I said. He's crossed the line one time too many, that one. And he's going to discover what we all realised a long time ago. That there isn't any way back.'

Acknowledgements

The usual list of suspects have contributed greatly to making this the book it is, and I unreservedly praise them all for their assistance whilst offering myself up for the inevitable critique for the bits of detail I've got wrong/writing style/too much bad language and so on.

My agent Sara O'Keeffe has been unfailingly helpful in guiding me through the sometimes choppy waters of writing fiction these days, and all that goes with it, and I am hugely grateful for her contributions to getting Mickey Bale onto the shelf rather than just living in the recesses of my fevered imagination. The team at Aevitas Creative have also proven to be strong allies, and my special thanks go to Syd James, Shenel Ekici-Moling and Allison Warren. And to Robin Wade, for seeing enough in the character to pass the baton to Sara in such a magnanimous style. I have truly been blessed twice with my choice of agents.

Editorially, Mickey has been well served, first by Holly Domney and latterly by Greg Rees, whose contributions have helped make *Target Zero* the strong book that I believe it to be. Helena Newton made a powerful contribution in the copy-edit, and since I believe that copy-editing tends to be an under-appreciated skill that can make a decent story so much better, I am very grateful. Speaking of gratitude, I also owe massive plaudits to Aries and Head of Zeus for stepping in where publishers are sometimes wary of treading these days, so thanks to Nic Cheetham for delivering Mickey kicking and punching into the world a second time.

As vitally as ever, I am constantly supported, frequently encouraged and occasionally (justifiably) chastised by my lovely wife Helen. Her constant stream of common sense, and her restraint of my occasional gnashing of teeth and garment rending (OK, daily) at the vicissitudes of modern life, writing, the day job and a dozen other subjects is as essential to my output as all the publishing expertise mentioned above. I am forever in her debt, and eternally grateful.

And lastly, yourselves. Thanks for reading the stories, and for the feedback of whatever nature that you provide online. Your support is hugely appreciated!

About the Author

ANTHONY RICHES, coming from a family with three generations of army service, has always been fascinated by military history, psychology and weaponry – which led him to write the Empire series set in ancient Rome. The idea for his first contemporary thriller, *Nemesis*, came to him under the influence of jetlag at two in the morning in a Brisbane hotel room, after a chance text discussion with a police officer. He lives in rural Suffolk with his wife, two dogs the size of ponies and a bad-tempered cat.

www.anthonyriches.com
@AnthonyRiches